Lecture Notes in Computer Science 7233

Commenced Publication in 1973
Founding and Former Series Editors:
Gerhard Goos, Juris Hartmanis, and Jan van Leeuwen

Andy Schürr   Dániel Varró
Gergely Varró (Eds.)

Applications of Graph Transformations with Industrial Relevance

4th International Symposium, AGTIVE 2011
Budapest, Hungary, October 4-7, 2011
Revised Selected and Invited Papers

 Springer

Volume Editors

Andy Schürr
Technische Universität Darmstadt, Real-Time Systems Lab
Merckstraße 25, 64283 Darmstadt, Germany
E-mail: andy.schuerr@es.tu-darmstadt.de

Dániel Varró
Budapest University of Technology and Economics
Department of Measurement and Information Systems
Magyar tudósok krt. 2, 1117 Budapest, Hungary
E-mail: varro@mit.bme.hu

Gergely Varró
Technische Universität Darmstadt, Real-Time Systems Lab
Merckstraße 25, 64283 Darmstadt, Germany
E-mail: gergely.varro@es.tu-darmstadt.de

ISSN 0302-9743   e-ISSN 1611-3349
ISBN 978-3-642-34175-5   e-ISBN 978-3-642-34176-2
DOI 10.1007/978-3-642-34176-2
Springer Heidelberg Dordrecht London New York

Library of Congress Control Number: 2012949456

CR Subject Classification (1998): G.2.2, E.1, D.2.1-2, D.2.4-5, D.2.11, F.2.2, F.3.1-2, F.4.2, I.2.8

LNCS Sublibrary: SL 2 – Programming and Software Engineering

Typesetting: Camera-ready by author, data conversion by Scientific Publishing Services, Chennai, India

Printed on acid-free paper

Springer is part of Springer Science+Business Media (www.springer.com)

Preface

This volume compiles all finally accepted papers presented at the 4th International Symposium on Applications of Graph Transformations with Industrial Relevance (AGTIVE 2011), which took place in October 2011 in Budapest, Hungary. The submissions underwent a thorough, two-round review process both before and after the symposium to enable the authors to carefully incorporate the suggestions of reviewers.

AGTIVE 2011 was the fourth practice-oriented scientific meeting of the graph transformation community. The aim of the AGTIVE series as a whole is to serve as a forum for all those scientists of the graph transformation community who are involved in the development of graph transformation tools and the application of graph transformation techniques usually in an industrial setting. In more detail, our intentions were to:

1. Bring the practice-oriented graph transformation community together
2. Study and integrate different graph transformation approaches
3. Build a bridge between academia and industry

The first AGTIVE symposium took place at Kerkrade, The Netherlands, in 1999. Its proceedings appeared as vol. 1779 of the Springer LNCS series. The second symposium was held in 2003 in Charlottesville, Virginia, USA. The proceedings appeared as LNCS vol. 3062. The third symposium took place in Kassel, Germany, in 2007, while the proceedings were published as LNCS vol. 5088.

AGTIVE 2011 was hosted by Budapest, the capital of Hungary, which was founded in 1873 as the unification of the separate historic towns of Buda (the royal capital since the fifteenth century), Pest (the cultural center), and Óbuda (built on the ancient Roman settlement of Aquincum). The city is bisected by the River Danube, which makes Budapest a natural geographical center and a major international transport hub. Budapest has a rich and fascinating history, a vibrant cultural heritage, yet it has managed to maintain its magic and charm. It has also been called the City of Spas with a dozen thermal bath complexes served by over a hundred natural thermal springs.

Thirty-six papers were submitted to AGTIVE 2011, which were evaluated twice by at least three reviewers. In all, 26 regular research papers, application reports, and tool demonstration papers were accepted in the first round for presentation at the symposium. These presentations covered a wide range of application areas such as model migration, software reengineering and configuration management, generation of test specifications, 3D reconstruction of plant architectures, chemical engineering, reconfiguration of self-adaptive systems, security aspects of embedded systems, and so forth. Furthermore, many contributions proposed new graph transformation concepts and implementation techniques needed to solve real-world problems in a scalable way.

In addition, two invited talks were presented at AGTIVE 2011 reporting on state-of-the-art, industrial rule-based modeling techniques and applications for business-critical applications.

- Zsolt Kocsis (IBM): "Best Practices to Model Business Services in Complex IT Environments"
- Mark Proctor (Red Hat) : "Drools: A Rule Engine for Complex Event Processing"

In the second review round, the Program Committee selected 18 submissions (of which 13 full research, two application report, and three tool demonstration papers) for publication in the proceedings. Thus, the final acceptance rate was 50%.

The AGTIVE 2011 Symposium was organized in strong collaboration between the Department of Measurement and Information Systems and the Department of Automation and Applied Informatics at the Budapest University of Technology and Economics, and the Real-Time Systems Lab at the Technical University Darmstadt. We would like to give special thanks to members of the organizing team, namely, Gábor Bergmann, Ábel Hegedüs, Ákos Horváth, László Lengyel, Gergely Mezei, István Ráth, Gábor Tóth, and Zoltán Ujhelyi. We also acknowledge the additional financial support provided by the SecureChange and the CERTIMOT projects, and the Alexander von Humboldt Foundation. The review process was administered by using the EasyChair conference management system.

July 2012

Andy Schürr
Dániel Varró
Gergely Varró

Organization

Program Co-chairs

Andy Schürr   TU Darmstadt, Germany
Dániel Varró   BME, Hungary
Gergely Varró   TU Darmstadt, Germany

Organizing Committee

Gábor Bergmann   Gergely Mezei
Ábel Hegedüs   István Ráth
Ákos Horváth   Gábor Tóth
László Lengyel   Zoltán Ujhelyi

Program Committee

Luciano Baresi   University of Milan, Italy
Benoit Baudry   INRIA, France
Paolo Bottoni   University of Rome La Sapienza, Italy
Jordi Cabot   INRIA, France
Krzysztof Czarnecki   University of Waterloo, Canada
Juan de Lara   Universidad Autónoma de Madrid, Spain
Hartmut Ehrig   Technical University of Berlin, Germany
Gregor Engels   University of Paderborn, Germany
Nate Foster   Cornell University, USA
Holger Giese   Hasso Plattner Institute, Germany
Reiko Heckel   University of Leicester, UK
Zhenjiang Hu   National Institute of Informatics, Japan
Audris Kalnins   University of Latvia, Latvia
Gabor Karsai   Vanderbilt University, USA
Ekkart Kindler   Technical University of Denmark, Denmark
Vinay Kulkarni   Tata Consultancy Services, India
Jochen Küster   IBM Research, Switzerland
Tihamér Levendovszky   Vanderbilt University, USA
Tom Mens   University of Mons-Hainaut, Belgium
Mark Minas   Universität der Bundeswehr München,
           Germany
Manfred Nagl   RWTH Aachen, Germany
Richard Paige   University of York, UK

Ivan Porres	Åbo Akademi University, Finland	
Arend Rensink	University of Twente, The Netherlands	
Leila Ribeiro	University of Rio Grande do Sul, Brazil	
Ingo Stürmer	Model Engineering Solutions GmbH, Germany	
Gabriele Taentzer	Philipps-Universität Marburg, Germany	
Pieter Van Gorp	Eindhoven University of Technology, The Netherlands	
Bernhard Westfechtel	University of Bayreuth, Germany	
Kang Zhang	University of Texas at Dallas, USA	
Albert Zündorf	Kassel University, Germany	

Additional Reviewers

Becker, Basil	Hegedüs, Ábel	Marchand, Jonathan
Branco, Moises	Hermann, Frank	Mosbah, Mohamed
Brüseke, Frank	Janssens, Dirk	Rutle, Adrian
Buchmann, Thomas	Jubeh, Ruben	Syriani, Eugene
Christ, Fabian	Koch, Andreas	Uhrig, Sabrina
Diskin, Zinovy	Krause, Christian	Winetzhammer, Sabine
Doberkat, Ernst-Erich	Lambers, Leen	
Geiger, Nina	Maier, Sonja	

Sponsoring Institutions

Department of Measurement and Information Systems, BME, Hungary
Department of Automation and Applied Informatics, BME, Hungary
Fachgebiet Echtzeitsysteme, Technische Universität Darmstadt, Germany
Alexander von Humboldt Foundation, Germany

Table of Contents

Invited Talk Abstracts

Session 1: Model-Driven Engineering

Session 2: Graph Transformation Applications

Session 3: Tool Demonstrations

Session 4: Graph Transformation Exploration Techniques

Session 5: Graph Transformation Semantics and Reasoning

Session 6: Application Reports

Session 7: Bidirectional Transformations

Best Practices to Model Business Services in Complex IT Environments

Zsolt Kocsis

IBM Hungary

Abstract. Managing complex business services on top of IT solutions is much more then managing solely the IT infrastructure beneath. Business services management requires the management of all resources and implementation layers by knowing the business logic, and includes services relevant information from different business aspects. To achieve an effective service modeling, the analysis must include resource models, connection models, error and error propagation models in a way that the models could be maintained to ensure long term business benefits.

This invited talk gives an insight of the business services modeling, a possible best practice to build such models in a complex, event based fault management environment and shows the actual outcomes of a recent project at a leading telecommunication company.

A. Schürr, D. Varró, and G. Varró (Eds.): AGTIVE 2011, LNCS 7233, p. 1, 2012.
© Springer-Verlag Berlin Heidelberg 2012

Drools: A Rule Engine
for Complex Event Processing

Mark Proctor

Red Hat

Abstract. Drools is the leading Java based Open Source rule engine. It is a hybrid chaining engine meaning it can react to changes in data and also provides advanced query capabilities. Drools provides built in temporal reasoning for complex event processing and is fully integrated with the jBPM project for BPMN2 based workflow. Ongoing research includes (but is not limited) to planning, ontological reasoning (semantic web), imperfect reasoning, truth maintenance and distributed collaboration through intelligent agents.

This talk will provide an introduction into Drools what it is and how it works. We will explain the concepts of forward and backward chaining within the context of Drools as well as exploring the rule engine syntax and how it has been extended for temporal reasoning for complex event processing.

A. Schürr, D. Varró, and G. Varró (Eds.): AGTIVE 2011, LNCS 7233, p. 2, 2012.

Graph Transformation Concepts for Meta-model Evolution Guaranteeing Permanent Type Conformance throughout Model Migration

Florian Mantz[1,*], Stefan Jurack[2], and Gabriele Taentzer[2]

[1] Bergen University College, Norway
fma@hib.no
[2] Philipps-Universität Marburg, Germany
{sjurack,taentzer}@mathematik.uni-marburg.de

Abstract. Meta-modeling has become the key technology to define domain-specific modeling languages for model-driven engineering. However, these modeling languages can change quite frequently which requires the evolution of their meta-models as well as the co-evolution (or migration) of their models. In this paper, we present an approach towards meta-model model co-evolution based on graph transformation concepts that targets to consider this challenge in a formal setting. Models are specified as graphs while model relations, especially type-instance relations, are defined by graph morphisms specifying type conformance of models to their meta-models. We present a basic approach to automatic deduction of model migrations from meta-model evolution steps which are specified by single transformation rules. Throughout that migration process, type conformance is ensured permanently. A first implementation is given using existing technology, namely the Eclipse Modeling Framework (EMF) and the EMF model transformation tool Henshin which is based on graph transformation concepts. Our evolution approach is presented at two small evolution scenarios for Petri nets and state machines.

Keywords: meta-model evolution, model migration, Henshin, graph transformation, Eclipse Modeling Framework.

1 Introduction

Model-driven engineering (MDE) is a software engineering discipline that uses models as main artifacts throughout software development processes, adopting model transformation for code generation. Models in MDE describe application-specific system design which is automatically translated into code. A commonly used technique to define modeling languages is meta-modeling.

In contrast to traditional software development where programming languages do not change often, domain-specific modeling languages and therefore meta-models, can change frequently: Modeling language elements are renamed, extended by additional attributes, or refined by a hierarchy of sub-elements. The evolution of a meta-model

* This work was partially funded by NFR project 194521 (FORMGRID).

A. Schürr, D. Varró, and G. Varró (Eds.): AGTIVE 2011, LNCS 7233, pp. 3–18, 2012.

requires the consistent migration of its models and is therefore a considerable research challenge in MDE (see Fig. 1).

Work in this direction already exists and several approaches have been developed to face this problem. Approaches use out-place and in-place transformation strategies which either create new migrated models like in [13] or change models in-place as in [6]. Most migration approaches are out-place. Out-place transformations have to translate all model parts, i.e., also those which do not change. In contrast, in-place transformations consider those model parts only that are affected by meta-model changes and can also support incremental migrations of large models. To the best of our knowledge, existing approaches do not ensure the permanent type conformance for in-place model migration processes. In general, existing approaches can be classified into the following categories: A model migration strategy is specified *manually* using a (specialized) model transformation language like in [13]. We consider a model migration to be *change-based*, i.e., if a difference model of two meta-model versions is calculated, an evolution rule is deduced and a migration definition like the one in [3] is automatically derived. Model migration is defined to be *operator-based*, if a set of coupled change strategies for meta-model evolution and model migration are provided to perform recurring adaptations like in [6].

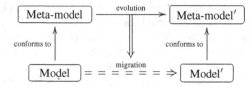

Fig. 1. Meta-model evolution and model migration

In this paper we explain how graph transformation concepts can be applied to meta-model evolution and model migration. Since graph transformations are in-place transformations, we migrate model parts only if they are affected by meta-model changes. In addition, we face the challenge of keeping models permanently type conformant to the evolving meta-model. We focus on an approach that is promising to formalize meta-model evolution with model migration while providing a prototypical implementation. The goal is to gain witness about completeness and correctness of model migration definitions before their application. Therefore we start with basic migration scenarios which will become more complex in the future. In addition, we develop an automatic derivation strategy for migration rules that does not rely on default migration operations. In our approach, we consider an in-place approach which uses an intermediate meta-model allowing the incremental transformation of large models. Using this migration strategy, models may also consist of many diagrams conforming to different meta-model versions. For example, consider a large UML 1.4 model consisting of many diagrams shared by several modelers. While some modelers are still working on their UML 1.4 diagrams others may already migrate their diagrams to a newer version of UML.

In this paper, models are consistently migrated using in-place transformations and a common meta-model. In contrast to our earlier work in [7], this coupled change operation can be generated from a simpler specification now, not only for one specific case but for a certain set of meta-model evolutions. Although there are several restrictions to evolution cases, the approach is already useful in various cases. Up to now, we study meta-model evolution steps which can be specified by single transformation rules without additional application conditions. Our migration strategy generates a migration rule r_M on the model level which is isomorphic to rule r_I evolving a corresponding meta-model. Rule r_M corresponds to rule r_I replacing each typing element in rule r_I by a corresponding instance element.

We realize our approach based on the Eclipse Modeling Framework (EMF) [5], a modeling framework widely used in practice. Ecore, the core language of EMF, complies with Essential MOF (EMOF) being part of OMG's Meta Object Facility (MOF) 2.0 specification [10]. The model transformation tool chosen to implement our approach is called Henshin [14]. It implements algebraic graph transformation concepts [4] adapted to EMF models and specifies its transformation systems as EMF models themselves. Henshin contains a control concept for rule applications called transformation units and furthermore, a Java(Script)-based computation engine for attributes. Since Henshin can be applied to any EMF model, we use it on two modeling levels: (1) to specify meta-model evolution rules and (2) to define a generator for migration rules.

This paper is structured as follows: In Section 2 we introduce our approach at a running example. In Sections 3 and 4, we explain the specification and derivation of meta-model evolution rules respectively model migration rules. Afterwards we discuss the implementation with EMF and Henshin in Section 5. Finally, in Sections 6 and 7, we consider related work and conclude our work.

2 Motivating Example

To motivate our approach, we consider two small evolution scenarios on Petri nets and state machines based on simple EMF-models (see Fig. 2). The essential part of the Ecore model is presented in Row 1 of Fig. 2. Its main elements are EClass and EReference. EReferences specify directed relations and their multiplicities. Furthermore they can be defined as containment references. For model storage, it is important that an EMF model forms a containment tree. For now, we neglect containment relationships here, but take them up again for the implementation. In addition, EClasses can have EAttributes typed over EDatatypes. Note that meta-references eStructuralFeatures and eType appear in the model twice, due to its flat presentation. In the original Ecore model, these meta-references are inherited from a common super class. Note also that EClass and EReference are objects in this "meta-meta-model". The Ecore model itself is recursively typed and EClass and EReference are both typed by EClass themselves.

Rows 2 and 4 of the example (Fig. 2) show meta-models in meta-model syntax. These meta-models are instances of our simplified Ecore meta-meta-model. Row 2 shows an evolution of a Petri net meta-model while Row 4 shows different versions of a state machine meta-model. Container classes are not shown due to space limitations.

However, it is always assumed that each model contains a root object with containment references to all other objects. The typing of meta-model elements is given by its syntax. Arrows are typed by `EReference`, nodes by `EClass`, attributes by `EAttribute` and data types by `EDatatype`.

Rows 3 and 5 of the example (Fig. 2) show instance models of the Petri net and state machine meta-models. The instance model typing is also given by its syntax. For example, place symbols in cell A3 are typed by `EClass` "Place" in cell A2. Note that the direction of incoming arrows in the Petri net meta-model in cell A2 does not correspond to the direction of their instance elements in cell A3. This is not a mistake but the meta-modeler's decision how to define the meta-model syntax of models. Examples of instance models in meta-model syntax are given in Fig. 3 which shows cells A3 and B3 of Fig. 2.

Rows 2 and 4 in Fig. 2 show two similar evolutions steps on two different meta-models. Rows 3 and 5 show their migrated models which may include additional values entered by the modeler. Cell A3 shows a simple Petri net which models a synchronous communication between "Alice" and "Bob". Cell B3 shows an extended version of the model. Weights at outgoing arrows are added allowing to express that "Alice" produces more messages than "Bob" during the communication. However, before the modeler can specify these weights it is required to evolve the meta-model and migrate the model. Afterwards he can change the default values. The outgoing arrow has to be specified as `EClass` in the meta-model since classes only can contain `EAttributes`. After replacing reference "outArr" by `EClass` "OutArr" and two additional references, an `EAttribute` "weight" for the weight at outgoing arrows can be added. If the modeler wants to use colored tokens in addition, the meta-model has to be evolved again. It is shown in cell C2 a meta-model with an additional attribute "color", a correspondingly migrated model in cell C3. The modeler transforms the `EAttribute` "token" of `EClass` "Place" into an own `EClass` where `EAttribute` "color" can be added. The old `EAttribute` "token" becomes the new `EAttribute` "count" in `EClass` "Token" which is connected by a new `EReference` with multiplicity 1:1.

All model migrations required should be implemented in a rule-based manner because a meta-model may have many instances which have to be migrated. It may also be an advantage to encode the meta-model evolution steps as rules. The evolution changes shown are quite general as we can see when we inspect the changes made to the state machine metamodel in Row 4 in Fig. 2. First, its instance models are extended by labeled transitions, then more information to actions is added such as the priority attribute "prio". Hence, we think that it is promising to encode meta-model evolution steps as reusable, parametrized transformation rules, applicable to any meta-model (see Figs. 4 and 5). Mappings are shown by numbers. Input parameter names start with "Var". Furthermore string manipulation functions can be used in rules. The rule in Fig. 4 replaces an `EReference` by an `EClass` linked by two `EReferences`. In addition, the new `EClass` gets an new `EAttribute` which is specified by the modeler applying this rule. The purpose of this evolution step is to prepare the meta-model for attributed links in the model.

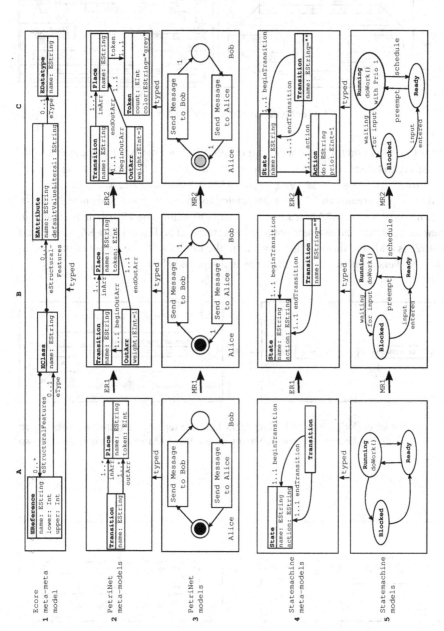

Fig. 2. Meta-model evolution and model migration

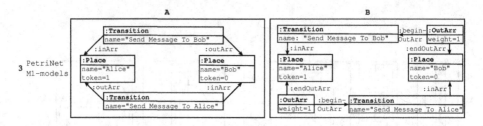

Fig. 3. Models in meta-model syntax

The rule application for the first change in cell A2 of our motivating example has the following arguments: `VarSrcNodeName="Transition"`, `VarArrName="out-Arr"`, `VarAttr="weight"`, `VarType="EInt"` and `VarAttrDefault="1"`. The rule in Fig. 5 creates an `EClass` associated to an existing one. This new `EClass` gets the name of a former `EAttribute`. To store its value a new attribute is created in this new `EClass` choosing a new name. Furthermore an additional new `EAttribute` "VarAttr2" is added to this new `EClass` as in the first rule. The purpose of this evolution step is to encapsulate several attributes in a separate class.

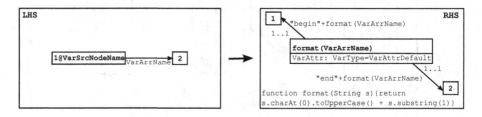

Fig. 4. Meta-model evolution Rule "EReference2EClass" for replacing an `EReference` by an `EClass`. In addition the new `EClass` contains a new `EAttribute` and is linked by two `EReferences`.

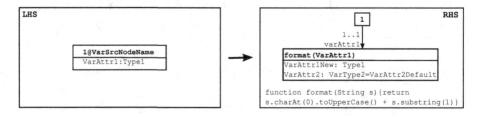

Fig. 5. Meta-model evolution Rule "EAttribute2AssociatedEClass" for replacing an attribute by an associated `EClass` containing the `EAttribute` renamed.

3   Meta-model Evolution

Since we want to use in-place transformations to perform the migration task we cannot apply the meta-model evolution rules directly. A direct application would destroy the typing morphisms as soon as we delete some of its elements. Hence, we have to do meta-model evolution and model migration in the following three steps (see Fig. 6): (1) extend the meta-model, (2) migrate all instance models, and (3) remove elements to be deleted from the meta-model. Following this process we can migrate models using an intermediate meta-model. The main advantage of this migration process is the permanent type conformance of instance models throughout their migration.

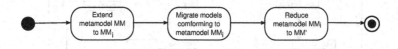

Fig. 6. Meta-model evolution, model migration process

To apply this process to meta-model evolution rules such as the ones in Figs. 4 and 5, we provide a transformation system (see Fig. 7) that generate the necessary artifacts for us. The transformation system derive the required meta-model evolution rules (to extend respectively restrict the metamodel) as well as model migration rule. This rule derivation is explained in the next sections.

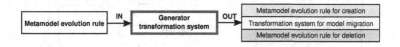

Fig. 7. Meta-model evolution rules

As mentioned before, meta-model evolution rules such as presented in Fig. 4 and Fig. 5 are not directly applied to evolve a meta-model in an in-place transformation approach. Rules like these two are called input rules since we use them to generate new rules to perform our migration approach.

As an example, a direct rule application of an input rule called "EReference2EClass" is illustrated in Fig. 8 in the usual meta-model syntax, while Fig. 9 shows this rule application in meta-meta-model syntax.

Since in our approach, rules follow algebraic graph transformation concepts, they can be formalized by using the well-known DPO-approach to graph transformation [4]. In [2], the theory (including node type inheritance) is enhanced with containment structures and constraints to capture all main EMF concepts. An abstract presentation of an input rule called r_I and its application is shown in Fig. 10. Models are specified by graphs and the type conformance of models to their meta-models is formalized by graph homomorphisms between instance and type graphs.

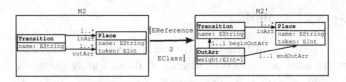

Fig. 8. Example meta-model evolution step: Before and after an application of Rule "ERefer-ence2EClass" in meta-model syntax

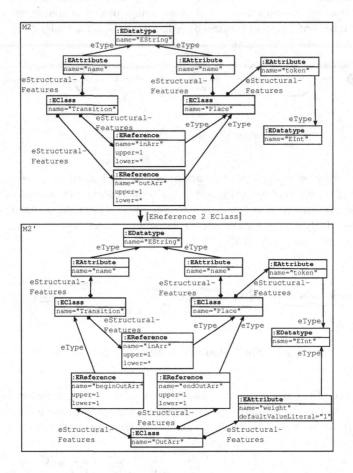

Fig. 9. Example meta-model evolution step: Before and after an application of Rule "ERefer-ence2EClass" in meta-meta-model syntax

As first step in our new approach, we split up rule r_I into two rules called r_C and r_D such that $[\![r_I]\!] = [\![r_C]\!]$, $[\![r_D]\!] = [\![r_C, r_D]\!]$ where $[\![r_I]\!]$ is the set of all possible applications of rule r_I. An illustration of this rule splitting is given in Fig. 11. Numbers in round brackets indicate the rule application order. Rule r_C creates elements only and rule r_D deletes elements only. Between the applications of these two rules we perform the

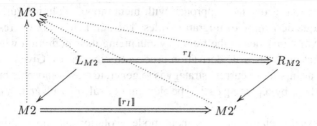

Fig. 10. Input rule r_I

migration transformation system which is considered in the next section. The LHS of rule r_C is equal to the LHS of rule r_I. The RHS of rule r_C as well as the LHS of rule r_D correspond to LHS $\cup$ RHS of rule r_I. The RHS of rule r_D is equal to the RHS of rule r_I.

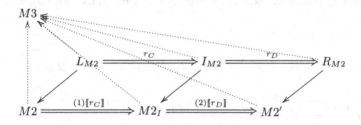

Fig. 11. Meta-model evolution rules

There are two general restrictions: Firstly, all changes must be expressible by "add" and "delete" operations i. e. an explicit move operation is not possible. This restriction is caused by the nature of graph transformation. Secondly, an intermediate meta-model containing all elements of the initial and revised meta-model must be constructible. This may not always be the case due to constraints of the meta-meta-model. For example, a meta-modeling language may prohibit multiple containers for the same object. However, multiple containment references may be required in the intermediate meta-model if the modeler wants to move an element from one container to another.

An example application of rules r_C and r_D is illustrated in the upper part of Fig. 13 for the concrete input rule "EReference2EClass". The application of rule r_C adds basically the structure EReference "beginOutArr", EClass "OutArr", EReference "endOutArr" to meta-model $M2$. The application of rule r_D deletes EReference "outArr" from the intermediate meta-model $M2_I$.

4   Model Migration

Models have to be migrated in a sensible way and may be done in more than one meaningful way. Hence often, this task cannot be fully automatized. For this reasons we start in this first version of our new approach with meta-model evolution rules that allow a sensible unique derivation of migration rules. The aim is to relax the restrictions on meta-model evolution rules in the future by enhancing their definition with additional information that guide the rule derivation process to other cases. Given a meta-model evolution rule as input, our current strategy is to generate one isomorphic model migration rule called r_M by replacing each type element t of rule r_I with an instance element of type t.

Fig. 12 shows the relation between meta-model evolution and model migration rules as well as their applications in a formal manner. The top of Fig. 12 shows the meta-model evolution rules r_C and r_D introduced in the previous section. Between their application, we execute the model migration rule r_M as often as possible. Rule r_C transforms type graph $M2$ into an intermediate type graph $M2_I$. An instance graph $M1$ is also type conformant to type graph $M2_I$ since $M2$ is included. Then, instance graph $M1$ is changed to instance graph $M1'$ by rule r_M. Rule r_M adds and deletes instance elements according to the input rule r_I. Afterwards, rule r_D is applied transforming type graph $M2_I$ into the final type graph $M2'$. Instance graph $M1'$ will also conform to this type graph since instances of types deleted by rule r_D have already been deleted by rule r_M before.

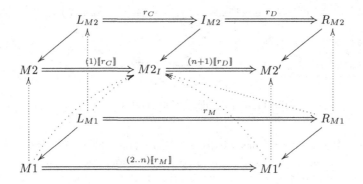

Fig. 12. Model migration transformation system

Currently, our deduction strategy for migration rules can be applied if meta-model evolution rules match graph parts only that cannot be varied on the instance level. Forbidden patterns are those that allow variance as e.g., loops and classes that are target of more than one reference. Whether a meta-model evolution rule belongs this class of rules can be checked in advance by analysing the meta-model evolution rule and constraints on the meta-model.

An example meta-model evolution with model migration by rule r_C, r_M and r_D is illustrated in Fig. 13. The process starts with applying rule r_C to meta-model $M2$ yielding $M2_I$. After this rule application, model $M1$ is still type conformant since nothing has been deleted in its meta-model $M2_I$. Then, the migration of model $M1$ follows by applying the derived rule r_M shown in Fig. 14 as often as possible. Finally the process ends with the application of rule r_D which deletes meta-model elements that are not used in the migrated model $M1'$ anymore since they have been replaced by other elements in the migration step before. Hence model $M1'$ is still type conformant. Note that rule r_M shown in Fig. 14 is presented in meta-model syntax. The typing is expressed using the usual ":"-notation. Compare also Fig. 14 with Fig. 4 to see that the structural parts of both rules are isomorphic. Note furthermore that the migration rule r_M does not require any variables anymore. The newly created EAttribute "weight" is initialized with the default value "1". However, the second meta-model evolution Rule "EAttribute2AssociatedEClass" requires an input parameter to transfer attribute values.

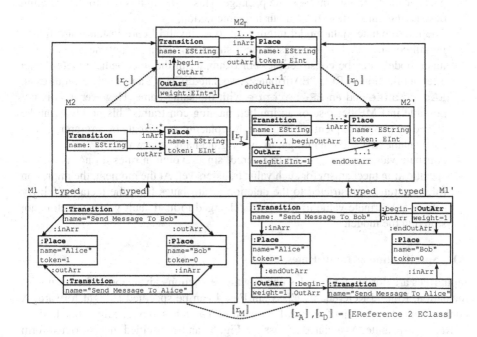

Fig. 13. Example meta-model evolution with model migration

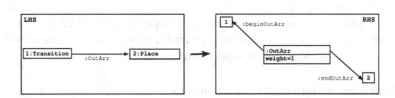

Fig. 14. Derived model migration rule for meta-model evolution Rule "EReference2EClass"

5   Implementation by Henshin Model Transformations

The prototypical implementation[1] is realized by Henshin model transformations. Henshin input rules are transformed to applicable meta-model evolution and model migration rules by use of higher-order transformations also specified with Henshin. Before we discuss the implementation realisation, we consider its current limitations:

1. The number of meta-model elements i.e., classes, references and attributes, to be deleted by the evolution rule is maximal one. This restriction is given only by the current implementation.
2. Newly created package elements have to be created within the context of the deleted package element, i.e., in the same package. This restriction is due to the handling of containment relationships in our implementation.
3. The intermediate meta-model might lead to an invalid Ecore instance not fulfilling all constraints. For this reason the restriction is by now that the intermediate meta-model must be constructible. For example, in the intermediate meta-model after application of Rule "EAttribute2AssociatedEClass", an EClass requires an EAttribute and an EReference with the same name. However, this is not possible in EMF because of an implicit naming constraint. This problem can be solved by two subsequent evolution steps replacing an attribute by a reference with a new name first and renaming it to the original name afterwards.
4. Attribute values cannot yet be transfered, since evolution rules do not provide a specification mechanism for such value transfers yet. At the moment, the migration rule generator is restricted to the deletion of attributes from the meta-model: By setting an attribute value to null it is changed to its default value, which is not stored in the model.

5.1   Specification of Input Rules

Input rules are normal Ecore rules obeying the restrictions given above. In our implementation, Rule "EReference2EClass" in Fig. 4 can be specified straight forward. A containment relationship of the newly created EClass has to be specified in addition.

Rule "EAttribute2Associated EClass" in Fig. 5 can be specified in a restricted form only. The value transfer of attribute "VarAttr1" cannot be specified, as already discussed. However, a specification mechanism for attribute value transfers is imaginable. In addition, we have to deal with another restriction: The name of the EReference to be create shall be specified by the name of an EAttribute. The intermediate meta-model would require that EClass "1" in the figure contains an EAttribute and an EReference with the same name. However, this is not possible in EMF. By now, we solve this problem by choosing another name for the EReference to create in a first step and adding a second step doing a renaming.

5.2   Derivation of Meta-model Evolution Rule

Our approach contains a model transformation system which creates two rules from scratch by the following coarse-grained algorithm: (1) Two rules r_C and r_D are created

---

[1] http://www.mathematik.uni-marburg.de/~swt/agtive-mm-evolution/

as copies of rule r_I. This is done by coping all elements, i.e., objects, links, mappings and parameters. (2) The RHS of rule r_C and the LHS of rule r_D are extended to $LHS_I \cup RHS_I$.

5.3 Derivation of Model Migration Rules

A model migration rule should perform corresponding actions on the model level as rule r_I does on the meta-model level. Therefore the creation of a migration rule is required resembling rule r_I adapted to the model level: Type elements of rule r_I have to be replaced by corresponding instance elements. Note that in EMF this translation is not that straight forward. Rule r_I is formulated in the meta-meta-model syntax, while rule r_M is specified in meta-model syntax: eReferences are objects in the meta-meta-model syntax and have to be translated into references in the meta-model syntax. Compare Fig. 3 with Fig. 9 to see which structures are mapped to each other. Fig. 9 shows a rule application of Rule "EReference2EClass" in meta-meta-model syntax. Fig. 3 shows a model in meta-model syntax before and after the meta-model change in Fig. 9. Furthermore, while the creation of an EAttribute does not need to be considered on the model level its deletion has to be treated in a special way: These attributes have to be set to null in the model as mentioned before (see Alg. 1). Newly created attributes exist implicitly in each model. Their default values can be specified in their meta-model. The move of attributes is not implemented yet. In addition, the creation of rule r_M requires that the elements of rule r_I can be used as types in migration rule r_M. Therefore we need an intermediate step: We generate a rule which generates the migration rule r_M for a specific meta-model. With this trick we can refer to the elements of rule r_I as types in this "generator generator" rule. However, we think that Henshin would profit from a feature supporting this step. In addition, the execution logic for applying migration rule r_M as often as possible has to be implemented with Henshin. This is done by generating not only the migration rule but a transformation system containing the migration rule and additional elements steering the rule application. If the evolution rule does creations only, the corresponding migration rule is applied exactly once. This procedure is formulated in the coarse-grained Algorithm 1 below.

5.4 Performance

The performance of this prototypical tool implementation is acceptable although there might be potentials for improvements. The derivation of one migration rule such as the one in Fig. 14 needs around 18 seconds on a standard laptop with an Intel(R) Core(TM)2 Duo CPU @ 2.40 GHz. The time spent to derive a migration rule was independent from the meta-model size in our test case: We applied rule "EReference2Class" to the Petri net and the State machine meta-models as well as to the Ecore meta-model. The performance of model migrations depend only on the used model transformation tool. We tested the derived model migration Rule "EReference2EClass" (see Fig. 14) with PetriNet models of different sizes. For a large PetriNet model with 1000 places and 1000 transitions where each one has one in and out arrow, our prototypical implementation needs around 240 seconds to complete the migration.

Algorithm 1. Calculate_model_migrator(Rule r_I) := TransformationSystem

---

1: TransformationSystem s := new HenshinTransformationSystem()
2: Rule r_M := $s.add(newRule())$
3: List E := All elements of LHS_{MM} and RHS_{MM} in rule r_I
4: **for** each element e in E **do**
5:     **if** e not instance of Attribute **then**
6:         Translate element e into its instance element i in r_M.
7:         **if** i instance of Node **then**
8:             Create a containment reference for i.
9:             Map corresponding instance node elements in LHS_M and RHS_M.
10:        **else**
11:            // i instance of Edge:
12:            Link the corresponding instance node elements in LHS_M resp. RHS_M.
13:        **end if**
14:    **else**
15:        **if** e is deleted by rule r_I /*Exist only in RHS_M*/ **then**
16:            // Delete attribute value in EMF instance:
17:            Translate element e into $e = null$ in RHS_M.
18:        **else**
19:            Do nothing. // New attributes exist implicitly in EMF models
20:        **end if**
21:    **end if**
22: **end for**
23: **if** rule r_M is not deleting **then**
24:     Embed rule r_M in s with a transformation unit calling rule r_M once.
25: **else**
26:     Embed rule r_M in s with transformation unit calling rule r_M as often as possible.
27: **end if**
28: **return** s

---

6   Related Work

Co-evolution of structures has been considered in several areas of computer science such as database schemata, grammars, and meta-models. Especially schema evolution has been a subject of research in the last decades. Recently, research activities have started to consider meta-model evolution and to investigate the transfer of schema evolution concepts to meta-model evolution. For further details we refer to Hermannsdörfer [6]. In the following, we focus our comparison on in-place and out-place approaches for meta-model evolution being presented in the literature.

In [13], Sprinkle et al. introduce a visual graph transformation-based language for meta-model evolution. The language uses out-place transformations and copies model elements automatically if their meta-model elements have not been changed. In contrast, an in-place approach does not need automatic copying of non-changed elements since a model can be changed directly.

In [9], Narayanan et al. introduce the Model Change Language (MCL) which mainly follows the work of [13]. Transformations are specified in a graph transformation based manner. The resulting model is checked for type conformance.

In [3], Cicchetti et al. present an changed-based approach. A difference model is automatically computed that acts as input for a higher-order transformation, producing a migration transformation. ATL [1] rules are generated. Helper classes are used to specify the transformation behavior.

Rose et al. in [12] introduce an automatic coping approach that uses an elaborated copy strategy. They present the text-based model migration language Epsilon Flock that can be used not only to copy unchanged model parts automatically but also those that pass a less strict conformance check. The idea is that a strict equivalence is often not required e.g., an integer object can be copied into a long object automatically. In addition, Epsilon Flock provides an abstraction mechanism to deal with different modeling frameworks.

In [6], the authors present a meta-model evolution approach called COPE that uses in-place transformations. A general difficulty of in-place transformations is that models and meta-models need to be changed in a coordinated manner as shown above, to guarantee a permanent conformance of models with their meta-model. However, COPE decouples models and meta-models during transformation and therefore cannot guarantee type conformance during model migration. Conformance is checked at runtime. A transaction concept comparable to the transaction concept in database systems is used to prevent models to get corrupted.

In [11] several tools for meta-model evolution are compared, namely: AtlanMod Matching Language (AML), Ecore2Ecore, COPE, Epsilon Flock. Except COPE, all tools in [11] use out-place transformations. AML is build on top of ATL and implements model matching strategies which execute a set of heuristics. Ecore2Ecore uses a mapping model to describe mappings between two Ecore models. Migration is specified by the use of this model and hand-written Java code.

In [8], Meyers et al. also suggest to use existing in-place transformation languages for model migration. Their approach mainly bases on the earlier work of Cicchetti et. al. in [3] but uses in-place transformations. It starts with a given set of difference models represented by a sequence of method calls that reflect a change in the meta-model and a model migration strategy. During the migration each model should stay conform to an intermediate meta-model as in our approach, however in their example implementation, they rely on COPE procedures.

7   Conclusion and Future Work

In this paper, an rule-based approach to meta-model evolution and model migration is presented using in-place model transformation based on graph transformation concepts. The approach is illustrated at two example evolutions of Petri net and state machine models showing that specific migration rules can be derived for different meta-models. The implementation is realized using existing technology, namely EMF and the EMF model transformation tool Henshin. Although our currently implemented migration strategy has a number of limitations, a variety of evolution case can be handled already now. In the future, the strategy for automatic deduction of migration rules has to be further elaborated to cover a larger set of meta-model evolution scenarios eliminating existing limitations step-by-step. Our intention is to develop the formal basis along with the elaboration of new strategies. The formalization of this work helps us to gain witness about the completeness and correctness of model migration.

References

1. Atlas Transformation Language: User Guide, http://wiki.eclipse.org/ATL/User_Guide
2. Biermann, E., Ermel, C., Taentzer, G.: Precise Semantics of EMF Model Transformations by Graph Transformation. In: Czarnecki, K., Ober, I., Bruel, J.-M., Uhl, A., Völter, M. (eds.) MODELS 2008. LNCS, vol. 5301, pp. 53–67. Springer, Heidelberg (2008)
3. Cicchetti, A., Di Ruscio, D., Eramo, R., Pierantonio, A.: Automating co-evolution in model-driven engineering. In: Proc. of the 12th International IEEE Enterprise Distributed Object Computing Conference, pp. 222–231. IEEE Computer Society (2008)
4. Ehrig, H., Ehrig, K., Prange, U., Taentzer, G.: Fundamentals of Algebraic Graph Transformation. Monographs in Theoretical Computer Science. An EATCS Series. Springer (2006)
5. EMF: Eclipse Modeling Framework (2010), http://www.eclipse.com/emf
6. Herrmannsdoerfer, M., Benz, S., Juergens, E.: COPE - Automating Coupled Evolution of Metamodels and Models. In: Drossopoulou, S. (ed.) ECOOP 2009. LNCS, vol. 5653, pp. 52–76. Springer, Heidelberg (2009)
7. Jurack, S., Mantz, F.: Towards metamodel evolution of EMF models with Henshin. Tech. rep., ME 2010: International Workshop on Model Evolution at MoDELS 2010 (Workshop Online Proceedings) (2010), http://www.modse.fr
8. Meyers, B., Wimmer, M., Cicchetti, A., Sprinkle, J.: A generic in-place transformation-based approach to structured model co-evolution. In: Amaral, V., Vangheluwe, H., Hardebolle, C., Lengyel, L. (eds.) Workshop on Multi-Paradigm Modeling 2010. ECEASST, vol. 42 (2011)
9. Narayanan, A., Levendovszky, T., Balasubramanian, D., Karsai, G.: Automatic Domain Model Migration to Manage Metamodel Evolution. In: Schürr, A., Selic, B. (eds.) MODELS 2009. LNCS, vol. 5795, pp. 706–711. Springer, Heidelberg (2009)
10. OMG: Meta-Object Facility 2.0. (2010), http://www.omg.org/spec/MOF/2.0/
11. Rose, L.M., Herrmannsdoerfer, M., Williams, J.R., Kolovos, D.S., Garcés, K., Paige, R.F., Polack, F.A.C.: A Comparison of Model Migration Tools. In: Petriu, D.C., Rouquette, N., Haugen, Ø. (eds.) MODELS 2010, Part I. LNCS, vol. 6394, pp. 61–75. Springer, Heidelberg (2010)
12. Rose, L.M., Kolovos, D.S., Paige, R.F., Polack, F.A.C.: Model Migration with Epsilon Flock. In: Tratt, L., Gogolla, M. (eds.) ICMT 2010. LNCS, vol. 6142, pp. 184–198. Springer, Heidelberg (2010)
13. Sprinkle, J., Karsai, G.: A domain-specific visual language for domain model evolution. Journal of Visual Languages and Computing 15(3-4) (2004)
14. The EMF Henshin Transformation Tool: Project Web Site, http://www.eclipse.org/modeling/emft/henshin/

A Graph Transformation-Based Semantics
for Deep Metamodelling

Alessandro Rossini[1], Juan de Lara[2], Esther Guerra[2], Adrian Rutle[3], and Yngve Lamo[3]

[1] University of Bergen, Norway
rossini@ii.uib.no
[2] Universidad Autónoma de Madrid, Spain
{Juan.deLara,Esther.Guerra}@uam.es
[3] Bergen University College, Norway
{aru,yla}@hib.no

Abstract. Metamodelling is one of the pillars of model-driven engineering, used for language engineering and domain modelling. Even though metamodelling is traditionally based on a two-level approach, several researchers have pointed out limitations of this solution and proposed an alternative *deep* (also called *multi-level*) approach to obtain simpler system descriptions. However, deep metamodelling currently lacks a formalisation that can be used to explain fundamental concepts such as deep characterisation through potency and double linguistic/ontological typing. This paper provides different semantics for such fundamental concepts based on graph transformation and the Diagram Predicate Framework.

1 Introduction

Model-driven engineering (MDE) promotes the use of models as the primary assets in software development, where they are used to specify, simulate, generate and maintain software systems. Models can be specified using general-purpose languages like UML, but to fully unfold the potential of MDE, models are specified using domain-specific languages (DSLs) which are tailored to a specific domain of concern. One way to define DSLs in MDE is by specifying metamodels, which are models that describe the concepts and define the syntax of a DSL.

The OMG has proposed MOF as the standard language to specify metamodels, and some popular implementations exist, most notably the Eclipse Modeling Framework (EMF) [21]. In this approach, a system is specified using models at two metalevels: a metamodel defining allowed types and a model instantiating these types. However, this approach may have limitations [4,5,13], in particular when the metamodel includes the *type-object* pattern [4,5,13], which requires an explicit modelling of types and their instances at the same metalevel. In this case, *deep metamodelling* (also called *multi-level metamodelling*) using more than two metalevels yields simpler models [5].

Deep metamodelling was proposed in the seminal works of Atkinson and Kühne [4], and several researchers and tools have subsequently adopted this approach [1,2,16]. However, there is still a lack of formalisation of the main concepts of deep metamodelling like deep characterisation through potency and double linguistic/ontological typing. Such a formalisation is needed in order to explain the main aspects of the approach,

A. Schürr, D. Varró, and G. Varró (Eds.): AGTIVE 2011, LNCS 7233, pp. 19–34, 2012.

study the different semantic variation points and their consequences, as well as to classify the different semantics found in the tools implementing them [1,2,3,16,15].

In this paper, we present a formal approach to deep metamodelling based on the Diagram Predicate Framework (DPF) [19,20], a diagrammatic specification framework founded on category theory and graph transformation. DPF has been adopted up to now to formalise several concepts in MDE, such as (MOF-based) metamodelling, model transformation and model versioning. The proposed formalisation helps in reasoning about the different semantic variation points in the realisation of deep metamodelling as well as in classifying the existing tools according to these options.

Paper Organisation. Section 2 introduces deep metamodelling through an example in the domain of component-based web applications. Section 3 presents the basic concepts of DPF. Section 4 explains different concepts of deep metamodelling through its formalisation in DPF. Section 5 compares with related research, and Section 6 concludes.

2   Deep Metamodelling

This section introduces deep metamodelling through an example, illustrating the limitations of two metalevels when defining DSLs which incorporate the type-object pattern [4,5,13]. Moreover, it discusses some open questions that are tackled in this paper.

2.1   Overview of Deep Metamodelling

The MeTEOriC project aims at the model-driven engineering of web applications. Here we describe a small excerpt of one of the modelling problems encountered in this project. A full description of this case study is outside the scope of this paper, but is described at: http://astreo.ii.uam.es/~jlara/metaDepth/Collab.html.

In MeTEOriC, a DSL is adopted to define the mash-up of components (like Google Maps and Google Fusion Tables) to provide the functionality of a web application. A simplified version of this language can be defined using two metalevels (see Fig. 1(a)). The metamodel corresponds to the DSL for component-based web applications. In this metamodel, the metaclass Component defines component types having a type identifier, whereas the metaclass CInstance defines component instances having a variable name and a flag indicating whether it should be visualised. Moreover, the metaassociation type defines the typing of each component instance. The model at the adjacent metalevel below represents a component-based web application which shows the position of professors' offices on a map. In this model, the classes Map and Table are instances of the metaclass Component and represent component types, whereas the classes UAM-Camp and UAMProfs are instances of the metaclass CInstance and represent component instances of Map and Table, respectively.

The type-object relation between component types and instances is represented explicitly in the metamodel by the metaassociation type between the metaclasses Component and CInstance. However, the type-object relation between allowed and actual data links is implicit since there is no explicit relation between the metaassociations datalink and dlinstance, and this may lead to several problems. Firstly, it is not possible to define that the data link instance offices is typed by the data link type geopos,

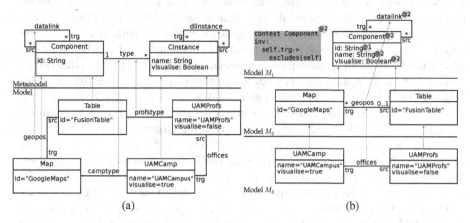

Fig. 1. A simple language for component systems in two and three metalevels

which could be particularly ambiguous if the model contained multiple data link types between the component types Map and Table. Moreover, it could be possible to specify a reflexive data link instance from the component instance UAMProfs to itself, which should not be allowed since the component type Table does not have any reflexive data link type. Although these errors could be detected by complementing the metamodel with attached OCL constraints, these constraints are not enough to guide the correct instantiation of each data link, in the same way as a built-in type system would do if the data link types and instances belonged to two different metalevels. In the complete definition of the DSL, the component types can define features, such as the zooming capabilities of the map component. Again, these features would be represented using the type-object pattern with metaclasses Feature (associated to Component) and FeatureInstance (associated to CInstance). These metaclasses need to be correctly instantiated and associated to the component instances, which leads to complex constraints and even more cluttered models. Hence, either one builds manually the needed machinery to emulate two metalevels within the same one, or this two-metalevel solution eventually becomes convoluted and hardly usable.

The DSL above can be defined in a simpler way using three metalevels (Fig. 1(b)) and deep characterisation, i.e., the ability to describe structure and express constraints for metalevels below the adjacent one. In this work, we adopt the deep characterisation approach described in [4], where each model element has a *potency*. In the original proposal of [4], the potency is a natural number which is attached to a model element to describe at how many subsequent metalevels this element can be instantiated. Moreover, the potency decreases in one unit at each instantiation at a deeper metalevel. When it reaches zero, a pure instance that cannot be instantiated further is obtained. In Section 4, we provide a more precise definition for potency.

In deep metamodelling, the elements in the top metalevel are pure types, the elements in the bottom metalevel are pure instances, and the elements at intermediate metalevels retain both a type and an instance facet. Because of that, they are all called *clabjects*, which is the merge of the words class and object [5]. Moreover, since in deep metamodelling the number of metalevels may change depending on the requirements, we find it more convenient to number the metalevels from 1 onwards starting from the top-most. The model M_1 contains the definition of the DSL (Fig. 1(b)). In this model, clabject Component has potency 2, denoting that it can be instantiated at the two subsequent metalevels. Its attribute id has potency 1, denoting that it can be assigned a value when Component is instantiated at the adjacent metalevel below. Its other two attributes name and visualise have potency 2, denoting that they can be assigned a value only two metalevels below. The association datalink also has potency 2, denoting that it can be instantiated at the two subsequent metalevels. The attached OCL constraint in the model M_1 forbids to reflexively connect indirect instances of Component. This constraint has potency 2, denoting that it has to be evaluated in the model M_3 only. As elements in M_2 retain a type facet, we can add cardinality constraints to geopos, while this would need to be emulated in Fig. 1(a). The DSL in Fig. 1(b) is simpler than the one in Fig. 1(a), as it contains less model elements to define the same DSL.

The deep characterisation is very useful in the design of this DSL. For instance, in the model M_1, the designer can specify the attributes name and visualise which should be assigned a value in indirect instances of Component, i.e., UAMCamp and UAMProfs. Moreover, the model M_1 does not need to include a clabject CInstance or an association dlinstance since the clabjects UAMCamp and UAMProfs are instances of the clabjects Map and Table, respectively, which in turn are instances of the clabject Component.

The dashed grey arrows in Fig. 1(b) denote the *ontological typing* for the clabjects, as they represent instantiations within a domain; e.g., the clabjects Map and Table are ontologically typed by the clabject Component. In addition, deep metamodelling frameworks usually support an orthogonal *linguistic typing* [5,16] which refers to the metamodel of the metamodelling language used to specify the models; e.g., the clabjects Component, Map and UAMCamp are linguistically typed by Clabject, whereas the attributes id, name and visualise are linguistically typed by Attribute.

2.2 Some Open Questions in Deep Metamodelling

Deep metamodelling allows a more flexible approach to metamodelling by introducing richer modelling mechanisms. However, their semantics have to be precisely defined in order to obtain sound, robust models. Even if the literature (and this section) permits grasping an intuition of how these modelling mechanisms work, there are still open questions which require clarification.

Some works in the literature give different semantics to the potency of associations. In Fig. 1(b), the associations are instantiated like clabjects. In this case, the association datalink with potency 2 in the model M_1 is first instantiated as the association geopos with potency 1 in the model M_2, and then instantiated as the association offices with potency 0 in the model M_3; i.e., the instantiation of offices is mediated by geopos. In contrast, the attributes name and visualise with potency 2 in the model M_1 are assigned a value directly in the model M_3; i.e., the instantiation of name and visualise is not mediated. Some frameworks such as EMF [21] represent associations as Java references, so the associations could also be instantiated like attributes. In this case, the association datalink would not need to be instantiated in the model M_2 in order to be able to instantiate it in the model M_3. This would have the effect that one could add an association between any two component instances in the model M_3, not necessarily between instances of Table and instances of Map.

Another ambiguity concerns constraints, since some works in the literature support potency on constraints [16] but others do not [3]. In Fig. 1(b), the attached OCL constraint in the model M_1 is evaluated in the model M_3 only. In other cases, it might be useful to have a potency which denotes that a constraint has to be evaluated at every metalevel. In addition, it is feasible to attach potencies to multiplicity constraints as well. In Fig. 1(b), all the multiplicity constraints are evaluated at the adjacent metalevel below. In other cases, it might be useful to attach a potency to multiplicity constraints. For instance, a potency 2 would have the effect that one could control the number of data link instances in the model M_3.

Finally, another research question concerns the relation between metamodelling stacks with and without deep characterisation. One could define constructions to *flatten* deep characterisation; e.g., given the three-metalevel stack of Fig. 1(b), one could obtain another three-metalevel stack without potencies but with some elements replicated along metalevels, making explicit the semantics of potency. This would allow the migration of deeply characterised systems into tools that do not support potency.

Altogether, we observe a lack of consensus and precise semantics for some of the aspects of deep metamodelling. The contribution of this work is the use of DPF to provide a neat semantics for different aspects of deep metamodelling: deep characterisation through potency and double linguistic/ontological typing. As a distinguishing note, we propose two possible semantics of potency for each model element, i.e., clabjects, attributes, associations and constraints. To the best of our knowledge, this is the first time that the two semantics have been recognised.

3 Diagram Predicate Framework

This section presents the basic concepts of DPF that are used in the formalisation of deep metamodelling. The interested reader can consult [9,8,10,18,19,17,20] for a more detailed presentation of the framework.

In DPF, a model is represented by a *specification* $\mathfrak{S}$. A specification $\mathfrak{S} = (S, C^{\mathfrak{S}} : \Sigma)$ consists of an *underlying graph* S together with a set of *atomic constraints* $C^{\mathfrak{S}}$ which are specified by means of a *predicate signature* Σ. A predicate signature $\Sigma = (\Pi^{\Sigma}, \alpha^{\Sigma})$ consists of a collection of *predicates* $\pi \in \Pi^{\Sigma}$, each having an arity (or shape graph) $\alpha^{\Sigma}(\pi)$. An atomic constraint (π, δ) consists of a predicate $\pi \in \Pi^{\Sigma}$ together with a graph homomorphism $\delta : \alpha^{\Sigma}(\pi) \to S$ from the arity of the predicate to the underlying graph of a specification.

Fig. 2 shows a specification $\mathfrak{T}$ which is compliant with the requirements "a component must have exactly one identifier", "a component may be connected to other components" and "a component can not be connected to itself". In $\mathfrak{T}$, these requirements are enforced by the atomic constraints ($[\mathtt{mult(1,1)}], \delta_1 : (1 \xrightarrow{a} 2) \to$ (Component $\xrightarrow{\text{id}}$ String)) and ($[\mathtt{irreflexive}], \delta_2 : (1 \xrightarrow{a} 1) \to$ (Component $\xrightarrow{\text{datalink}}$ Component)).

Similar to E-graphs [11], attributes of nodes can be represented in DPF by edges from these nodes to nodes representing data types. For example, the attribute id:String of the clabject Component in Fig. 1(b) is represented in DPF by an edge Component $\xrightarrow{\text{id}}$ String (see Fig. 2).

The semantics of graph nodes and arrows has to be chosen in a suitable way for the corresponding modelling environment [20]. In object-oriented structural modelling, each object may be related to a set of other objects. Hence, it is appropriate to interpret nodes as sets and arrows $X \xrightarrow{f} Y$ as multi-valued functions $f : X \to \wp(Y)$.

The semantics of predicates of the signature Σ (see Fig 2) is described using the mathematical language of set theory. In an implementation, the semantics of a predicate is typically given by the code of a corresponding validator such that the mathematical and the validator semantics should coincide. A semantic interpretation $[\![..]\!]^{\Sigma}$ of a signature Σ consists of a mapping that assigns to each predicate symbol $\pi \in \Pi^{\Sigma}$ a set $[\![\pi]\!]^{\Sigma}$ of graph homomorphisms $\iota : O \to \alpha^{\Sigma}(\pi)$, called valid instances of π, where O may vary over all graphs. $[\![\pi]\!]^{\Sigma}$ is assumed to be closed under isomorphisms.

The semantics of a specification is defined in the fibred way [8,10]; i.e., the semantics of a specification $\mathfrak{S} = (S, C^{\mathfrak{S}} : \Sigma)$ is given by the set of its instances (I, ι). To check that an atomic constraint is satisfied in a given instance of a specification $\mathfrak{S}$, it is enough to inspect only the part of $\mathfrak{S}$ which is affected by the atomic constraint. This kind of restriction to a subpart is obtained by the pullback construction [6]. An instance (I, ι) of a specification $\mathfrak{S}$ consists of a graph I and a graph homomorphism $\iota : I \to S$ such that for each atomic constraint $(\pi, \delta) \in C^{\mathfrak{S}}$ we have $\iota^* \in [\![\pi]\!]^{\Sigma}$, where the graph homomorphism $\iota^* : O^* \to \alpha^{\Sigma}(\pi)$ is given by the following pullback:

$$
\begin{array}{ccc}
\alpha^{\Sigma}(\pi) & \xrightarrow{\delta} & S \\
{\scriptstyle \iota^*}\Big\uparrow & P.B. & \Big\uparrow{\scriptstyle \iota} \\
O^* & \xrightarrow{\delta^*} & I
\end{array}
$$

In DPF, two kinds of conformance relations are distinguished: *typed by* and *conforms to*. A specification $\mathfrak{S}$ is typed by a graph T if there exists a graph homomorphism $\iota : S \to T$, called the *typing morphism*, between the underlying graph of the specification $\mathfrak{S}$ and T. A specification $\mathfrak{S}$ is said to conform to a specification $\mathfrak{T}$ if there exists a typing morphism $\iota : S \to T$ between the underlying graphs of $\mathfrak{S}$ and $\mathfrak{T}$ such that (S, ι) is a valid instance of $\mathfrak{T}$; i.e., such that ι satisfies the atomic constraints $C^{\mathfrak{T}}$.

Fig. 2 shows two specifications $\mathfrak{S}$ and $\mathfrak{S}'$, both typed by $\mathfrak{T}$. However, only $\mathfrak{S}$ conforms to $\mathfrak{T}$, since $\mathfrak{S}'$ violates the atomic constraints $C^{\mathfrak{T}}$. This is because the missing id-typed edge violates the multiplicity constraint ([mult(1,1)], δ_1), while the edge source violates the irreflexivity constraint ([irreflexive], δ_2).

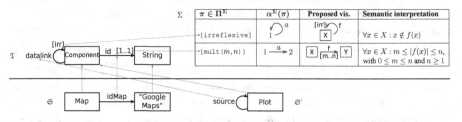

Σ	$\pi \in \Pi^{\Sigma}$	$\alpha^{\Sigma}(\pi)$	Proposed vis.	Semantic interpretation
	[irreflexive]	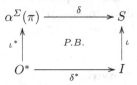		$\forall x \in X : x \notin f(x)$
	[mult(m,n)]	$1 \xrightarrow{a} 2$		$\forall x \in X : m \le |f(x)| \le n$, with $0 \le m \le n$ and $n \ge 1$

Fig. 2. A signature Σ and specifications $\mathfrak{T}$, $\mathfrak{S}$ and $\mathfrak{S}'$, where only $\mathfrak{S}$ conforms to $\mathfrak{T}$

4 Formalisation of Deep Metamodelling

This section formalises different concepts of deep metamodelling through DPF. Firstly, we introduce different interpretations of potency. Secondly, we define the syntax of potency in terms of DPF. Thirdly, we define models in a deep stack together with double linguistic/ontological typing in terms of DPF. Finally, we present an operational semantics of potency in terms of constraint-aware graph transformation.

4.1 Multi- and Single-Potency

As discussed in Section 2.2, different interpretations of potency are possible. In this paper, two kinds of potency are distinguished, namely *multi-* and *single*-potency, denoted by the superscripts $\blacktriangle$p and $\triangle$p, respectively.

A multi-potency $\blacktriangle$p on a clabject/reference at metalevel i denotes that this clabject/reference can be instantiated *at all metalevels from $i + 1$ to $i + p$* (see Fig. 3). A potency $\blacktriangle$p on an atomic constraint at metalevel i denotes that this constraint is evaluated at all metalevels from $i + 1$ to $i + p$. Note that attributes only retain either type or instance facet but not both, therefore the multi-potency on attributes can not be considered. This "multi-" semantics is the usual semantics of potency on clabjects found in the literature.

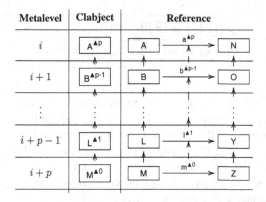

Fig. 3. Intuition on the multi-semantics of potency

In contrast, a single-potency $\triangle$p on a clabject/reference at metalevel i denotes that this clabject/reference can be instantiated *at metalevel $i + p$ only*, but not at the intermediate metalevels (see Fig. 4). A potency $\triangle$p on an attribute at metalevel i denotes that this attribute can be instantiated (i.e., can be assigned a value) at metalevel $i + p$ only. A potency $\triangle$p on an atomic constraint at metalevel i denotes that this atomic constraint is evaluated at metalevel $i + p$ only.

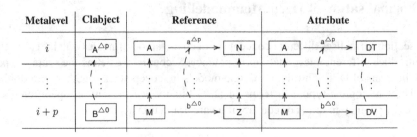

Fig. 4. Intuition on the single- semantics of potency

4.2 Syntax of Potency

The syntax of multi- and single-potencies can be represented in DPF by a *tag signature*, which has the same components of a predicate signature but a different semantic counterpart (see Section 4.3 and 4.4). A tag signature $\Psi = (\Theta^{\Psi}, \alpha^{\Psi})$ consists of a collection of *tags* $\theta \in \Theta^{\Psi}$, each having an arity $\alpha^{\Psi}(\theta)$ and a proposed visualisation. Table 1 shows the tag signature Ψ for specifying potencies.

Table 1. The tag signature Ψ for specifying potencies

$\theta \in \Theta^{\Psi}$	$\alpha^{\Psi}(\theta)$	Proposed visual.	$\theta \in \Theta^{\Psi}$	$\alpha^{\Psi}(\theta)$	Proposed visual.
`<multi(p)>`^{1}	1	$X^{\blacktriangle p}$	`<single(p)>`^{1}	1	$X^{\triangle p}$
`<multi(p)>`^{2}	$1 \xrightarrow{a} 2$	$X \xrightarrow{f^{\blacktriangle p}} Y$	`<single(p)>`^{2}	$1 \xrightarrow{a} 2$	$X \xrightarrow{f^{\triangle p}} Y$
`<multi(p)>`$^\pi$	$1 \xrightarrow{a} 2$	$X \xrightarrow[\pi^{\blacktriangle p}]{f} Y$	`<single(p)>`$^\pi$	$1 \xrightarrow{a} 2$	$X \xrightarrow[\pi^{\triangle p}]{f} Y$

The tags $\theta \in \Theta^{\Psi}$ are divided into two families `<multi(p)>` and `<single(p)>` for multi- and single-potency, respectively. They are parametrised by the (non-negative) integer p, which represents the potency value that is attached to an element. More specifically, the tags `<multi(p)>`^{1} and `<single(p)>`^{1} are used for attaching potencies to clabjects, `<multi(p)>`^{2} and `<single(p)>`^{2} for references and attributes, and `<multi(p)>`$^\pi$ and `<single(p)>`$^\pi$ for atomic constraints (with compatible arity).

Given the tag signature Ψ, a potency (θ, γ) consists of a tag $\theta \in \Theta^{\Psi}$ and a graph homomorphism $\gamma : \alpha^{\Psi}(\theta) \to S_i$. Note that potencies can only be attached to clabjects, references, attributes and atomic constraints. This restriction can be defined by adopting typed tag signatures in which each tag is typed linguistically by a specification. This detail is omitted in this paper for brevity.

In the following, we adopt potencies to define models in a deep stack.

4.3 Double Linguistic/Ontological Typing

A model at metalevel i in a deep stack can be represented in DPF by a *(deep) specification* $\mathfrak{S}_i$. A specification $(\mathfrak{S}_i, \lambda_i, \omega_i) = (S_i, C_i : \Sigma, P_i : \Psi, \lambda_i, \omega_i)$ consists of an underlying graph S_i, a set of atomic constraints C_i specified by means of a predicate signature Σ and a set of potencies P_i specified by means of a tag signature Ψ (see Fig. 5). Moreover, $\mathfrak{S}_i$ conforms linguistically to the specification $\mathfrak{LM}$; i.e., there exists a linguistic typing morphism $\lambda_i : S_i \to LM$ such that (S_i, λ_i) is a valid instance of $\mathfrak{LM}$. The specification $\mathfrak{LM}$ corresponds to the metamodelling language used to specify all specifications in a deep stack. Furthermore, $\mathfrak{S}_i$ conforms ontologically to the specification $\mathfrak{S}_{i-1}$; i.e., there exists an ontological typing morphism $\omega_i : S_i \to S_{i-1}$ such that (S_i, ω_i) is a valid instance of $\mathfrak{S}_{i-1}$ and the ontological typing is compatible with the linguistic typing, i.e., $\omega_i; \lambda_{i-1} = \lambda_i$.

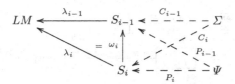

Fig. 5. Double linguistic/ontological typing

A specification $\mathfrak{S}_1$ at the top metalevel 1 of a deep stack is a special case as its elements are pure types (see Section 2). As such, $\mathfrak{S}_1$ conforms only linguistically to the specification $\mathfrak{LM}$; i.e., there is no specification $\mathfrak{S}_0$, hence, there is no ontological typing morphism $\omega_1 : S_1 \to S_0$.

Example 1 (Deep Stack). Building on the example in Section 2, Fig. 6(a) shows the specification $\mathfrak{LM}$ while Figs. 6(b), (c) and (d) show the specifications $\mathfrak{S}_1$, $\mathfrak{S}_2$ and $\mathfrak{S}_3$ of a deep stack corresponding to a simplified version of the one in Fig. 1(b). The figure also shows the ontological typings as dashed, grey arrows.

In $\mathfrak{S}_1$ the potency ▲2 on Component and datalink is specified by $(\texttt{<multi(2)>}^1$,$\gamma_1 : 1 \to$ Component$)$ and $(\texttt{<multi(2)>}^2, \gamma_2 : (1 \xrightarrow{a} 2) \to$ (Component $\xrightarrow{\text{datalink}}$ Component)), respectively. Similarly, the potencies △1 on id and $(\texttt{[mult(1,1)]}, \delta_1)$ are specified by $(\texttt{<single(1)>}^2, \gamma_3)$ and $(\texttt{<single(1)>}^2, \gamma_4; \delta_1)$, respectively.

The specifications $\mathfrak{S}_1$, $\mathfrak{S}_2$ and $\mathfrak{S}_3$ conform linguistically to $\mathfrak{LM}$; i.e., there exist linguistic typing morphisms $\lambda_1 : S_1 \to LM$, $\lambda_2 : S_2 \to LM$ and $\lambda_3 : S_3 \to LM$ such that (S_1, λ_1), (S_2, λ_2) and (S_3, λ_3) are valid instances of $\mathfrak{LM}$. For instance, λ_2 is defined as follows:

$\lambda_2(\text{Map}) = \lambda_2(\text{Table}) = \text{Clabject}$
$\lambda_2(\text{geopos}) = \text{Reference}$
$\lambda_2(\text{idMap}) = \lambda_2(\text{idTable}) = \text{Attribute}$
$\lambda_2(\text{“GoogleMaps”}) = \lambda_2(\text{“FusionTable”}) = \text{DataType}$

Moreover, $\mathfrak{S}_2$ conforms ontologically to $\mathfrak{S}_1$; i.e., there exists an ontological typing morphism $\omega_2 : S_2 \to S_1$ such that (S_2, ω_2) is a valid instance of $\mathfrak{S}_1$:

$\omega_2(\text{Map}) = \omega_2(\text{Table}) = \text{Component}$
$\omega_2(\text{geopos}) = \text{datalink}$
$\omega_2(\text{idMap}) = \omega_2(\text{idTable}) = \text{id}$
$\omega_2(\text{“GoogleMaps”}) = \omega_2(\text{“FusionTable”}) = \text{String}$

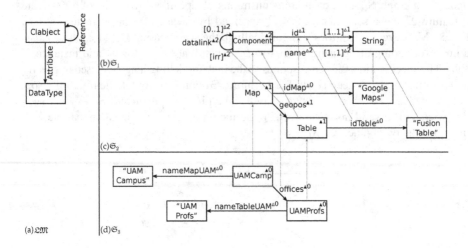

Fig. 6. The specifications $\mathfrak{LM}$, $\mathfrak{S}_1$, $\mathfrak{S}_2$ and $\mathfrak{S}_3$

Finally, $\mathfrak{S}_3$ should conform ontologically to $\mathfrak{S}_2$, but this is not the case as the ontological typing morphism ω_3 is undefined for some elements of $\mathfrak{S}_3$:

$\omega_3(\mathsf{UAMCamp}) = \mathsf{Map}$
$\omega_3(\mathsf{UAMProfs}) = \mathsf{Table}$
$\omega_3(\mathsf{offices}) = \mathsf{geopos}$
$\omega_3(\mathsf{nameMapUAM}) = \omega_3(\mathsf{nameTableUAM}) = \omega_3(\text{"UAMCampus"}) = \omega_3(\text{"UAMProfs"}) = \emptyset$

In the following, we adopt constraint-aware graph transformation to define an operational semantics of single-potency and obtain a specification $\mathfrak{S}_3$ which conforms ontologically to $\mathfrak{S}_2$.

4.4 Semantics of Potency through Graph Transformation

Recall that a single-potency $\triangle p$ on a type at metalevel i denotes that this type can be instantiated *at metalevel $i + p$ only*. Hence, there are always p metalevels between an instance with potency 0 and its type. However, in strict metamodelling, an instance with potency 0 at metalevel $i + p$ should be ontologically typed by a type with potency 1 at metalevel $i + p - 1$. To address this problem we define a semantics of single-potency which transforms a deep stack into a flattened stack without deep characterisation, in which an instance with potency 0 at metalevel $i+p$ has its type at metalevel $i+p-1$. This transformation adds to each metalevel $i + 1$ a replica of a type with potency decreased to $p-1$ and then deletes from metalevel i the original type, until $p = 1$. We specify this transformation with constraint-aware *transformation rules* [11].

A transformation rule $t = \mathfrak{L} \xleftarrow{\ l\ } \mathfrak{K} \xrightarrow{\ r\ } \mathfrak{R}$ consists of three specifications $\mathfrak{L}$, $\mathfrak{K}$ and $\mathfrak{R}$. $\mathfrak{L}$ and $\mathfrak{R}$ are the *left-hand side* (LHS) and *right-hand side* (RHS) of the transformation rule, respectively, while $\mathfrak{K}$ is their interface. $\mathfrak{L} \setminus \mathfrak{K}$ describes the part of a specification which is to be deleted, $\mathfrak{R} \setminus \mathfrak{K}$ describes the part to be added and $\mathfrak{K}$ describes the part to be preserved by the rule. Roughly speaking, an *application of transformation rule* means finding a match of the left-hand side $\mathfrak{L}$ in a source specification $\mathfrak{S}$ and replacing $\mathfrak{L}$ by $\mathfrak{R}$, leading to a target specification $\mathfrak{S}'$.

Since the transformation a specification undergoes is dependent of the potencies in the specification at the metalevel above, the transformation rules take as input and output *coupled specifications*. A coupled specification $\mathfrak{CS}_i = ((\mathfrak{S}_i, \lambda_i, \omega_i), (\mathfrak{S}_{i+1}, \lambda_{i+1}))$ consists of a specification $(\mathfrak{S}_i, \lambda_i, \omega_i)$ coupled with the specification $(\mathfrak{S}_{i+1}, \lambda_{i+1})$. This means that the application of transformation rules modifies specifications at two adjacent metalevels i and $i + 1$.

Table 2 shows some of the transformation rules which define the operational semantics of single-potency. In general, all these rules follow a general pattern which adds to metalevel $i + 1$ a replica of an element with single-potency decreased to $p - 1$ and then deletes from metalevel i the original element; i.e.:

- Rules tc_0 and tdt_0: add to metalevel $i + 1$ a replica of a clabject/data type.
- Rules tr_1 and ta_1: add to metalevel $i + 1$ a replica of a reference/attribute.
- Rules $tacr_2$ and $taca_2$: add to metalevel $i + 1$ a replica of an atomic constraint.
- Rules $tacr_3$ and $taca_3$: delete from metalevel i the original atomic constraint.
- Rules tr_4 and ta_4: delete from metalevel i the original reference/attribute.
- Rule tc_5: deletes from metalevel i the original clabject.

Note that the rules tc_0, tc_5, $taca_2$ and $taca_3$ are analogous to the rules tr_1, tr_4, $tacr_2$ and $tacr_3$, respectively, and are omitted from Table 2 for brevity.

The transformation uses negative application conditions (NACs) and layers [11] to control rule application. Since non-deleting rules can be applied multiple times via the same match, each rule has a NAC equal to its RHS. Moreover, since the rules are to be applied only if the matched potency is greater than 1, rules have another NAC demanding $p \leq 1$. The subscripts from 0 to 4 denote the layer to which a rule belongs, so that rules of layer 0 are applied as long as possible before rules of layer 1, etc.

According to this layering, the transformation adds a replica of a reference only *after* it adds a replica of a clabject and *before* it deletes the original clabject. This ensures that the rule which adds a replica of a reference matches both clabjects with multi-potency and their instances as well as clabjects with single-potency and their replicas. Moreover, this ensures that the replica of the reference has as source and target an instance of the considered clabject with multi-potency or a replica of the considered clabject with single-potency. The layering of rules for data types, attributes and atomic constraints follow the same rationale.

Note that the notation $\boxed{\text{A:Clabject}} \xrightarrow{\text{a:Attribute}} \boxed{\text{DT:DataType}}$ in the rules denotes that $\lambda_i(A) = \text{Clabject}$, $\lambda_i(a) = \text{Attribute}$ and $\lambda_i(DT) = \text{DataType}$.

Example 2 (Deep Stack and Application of Transformation Rules). Building on Example 1, Figs. 7(b), (c) and (d) show the specifications $\mathfrak{S}_1$, $\mathfrak{S}_2$ and $\mathfrak{S}_3$ of the deep stack, after the application of the transformation rules. In particular, the added elements are shown in Fig. 7(c) in green colour while the deleted elements are shown in Fig. 7(b) in red colour.

Table 2. The transformation rules for flattening the semantics of single-potencies

	$\mathfrak{CL} = \mathfrak{CR}$	$\mathfrak{CR} = \mathfrak{NAC}$
tdt_0	A:Clabject $\xrightarrow{\text{a:Attribute}^{\triangle p}}$ DT:DataType; ω_i^L : B:Clabject	A:Clabject $\xrightarrow{\text{a:Attribute}^{\triangle p}}$ DT:DataType; ω_i^R : B:Clabject — ω_i^R : DT:DataType
ta_1	A:Clabject $\xrightarrow{\text{a:Attribute}^{\triangle p}}$ DT:DataType; ω_i^L ; B:Clabject ⋯ ω_i^L ⋯ DT:DataType	A:Clabject $\xrightarrow{\text{a:Attribute}^{\triangle p}}$ DT:DataType; ω_i^R ; B:Clabject $\xrightarrow{\text{aB:Attribute}^{\triangle p\text{-}1}}$ DT:DataType; ω_i^R
tr_1	A:Clabject $\xrightarrow{\text{a:Reference}^{\triangle p}}$ N:Clabject; ω_i^L ; B:Clabject ⋯ ω_i^L ⋯ O:Clabject	A:Clabject $\xrightarrow{\text{a:Reference}^{\triangle p}}$ N:Clabject; ω_i^R ; B:Clabject $\xrightarrow{\text{aB:Reference}^{\triangle p\text{-}1}}$ O:Clabject; ω_i^R
$tacr_2$	A:Clabject $\xrightarrow[\text{a:Reference}]{\pi^{\triangle p}}$ N:Clabject; ω_i^L ; B:Clabject $\xrightarrow{\text{aB:Reference}}$ O:Clabject; ω_i^L	A:Clabject $\xrightarrow[\text{a:Reference}]{\pi^{\triangle p}}$ N:Clabject; ω_i^R ; B:Clabject $\xrightarrow[\text{aB:Reference}]{\pi^{\triangle p\text{-}1}}$ O:Clabject; ω_i^R

	$\mathfrak{CL}$	$\mathfrak{CK} = \mathfrak{CR}$
$tacr_3$	A:Clabject $\xrightarrow[\text{a:Reference}]{\pi^{\triangle p}}$ N:Clabject; ω_i^L ; B:Clabject $\xrightarrow[\text{aB:Reference}]{\pi^{\triangle p\text{-}1}}$ O:Clabject; ω_i^L	A:Clabject $\xrightarrow[\text{a:Reference}]{\pi^{\triangle p}}$ N:Clabject; ω_i^R ; B:Clabject $\xrightarrow[\text{aB:Reference}]{\pi^{\triangle p\text{-}1}}$ O:Clabject; ω_i^R
tr_4	A:Clabject $\xrightarrow{\text{a:Reference}^{\triangle p}}$ N:Clabject; ω_i^L ; B:Clabject $\xrightarrow{\text{aB:Reference}^{\triangle p\text{-}1}}$ O:Clabject; ω_i^L	A:Clabject N:Clabject; ω_i^R ; B:Clabject $\xrightarrow{\text{aB:Reference}^{\triangle p\text{-}1}}$ O:Clabject; ω_i^R
ta_4	A:Clabject $\xrightarrow{\text{a:Attribute}^{\triangle p}}$ DT:DataType; ω_i^L ; B:Clabject $\xrightarrow{\text{aB:Attribute}^{\triangle p\text{-}1}}$ DT:DataType; ω_i^L	A:Clabject DT:DataType; ω_i^R ; B:Clabject $\xrightarrow{\text{aB:Attribute}^{\triangle p\text{-}1}}$ DT:DataType; ω_i^R

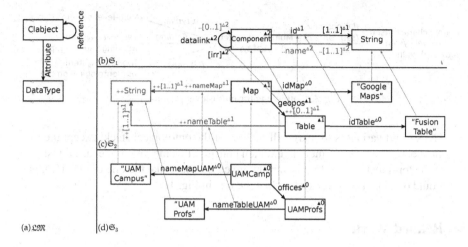

Fig. 7. The specifications $\mathfrak{S}_1$, $\mathfrak{S}_2$ and $\mathfrak{S}_3$, after applying the rules

Firstly, the application of tdt_0 and ta_1 adds to $\mathfrak{S}_2$ the node String and the edges nameMap and nameTable with potency $\triangle 1$. In this way, the ontological typing morphism ω_3 can be defined for all the elements of $\mathfrak{S}_3$, which makes $\mathfrak{S}_3$ conform ontologically to $\mathfrak{S}_2$:

$$\omega_3(\text{nameMapUAM}) = \text{nameMap}$$
$$\omega_3(\text{nameTableUAM}) = \text{nameTable}$$
$$\omega_3(\text{"UAMCampus"}) = \omega_3(\text{"UAMProfs"}) = \text{String}$$

Secondly, the application of $tacr_2$ and $taca_2$ adds to $\mathfrak{S}_2$ the atomic constraints $([\texttt{mult}(0,1)],\delta_1)$, $([\texttt{mult}(1,1)],\delta_2)$ and $([\texttt{mult}(1,1)],\delta_3)$ with potency $\triangle 1$ on the edges geopos, nameMap and nameTable, respectively. In this way, these atomic constraints are evaluated at metalevel 0.

Thirdly, the application of $tacr_3$ and $taca_3$ deletes from $\mathfrak{S}_1$ the atomic constraints $([\texttt{mult}(0,1)],\delta_1)$ and $([\texttt{mult}(1,1)],\delta_3)$ on the edges datalink and name, respectively. In this way, these atomic constraints are not evaluated at metalevel 1.

Finally, the application of $taca_4$ deletes from $\mathfrak{S}_1$ the edge name. In this way, it is not possible to instantiate name at metalevel 1.

The presented flattening gives the semantics of potencies, so that an equivalent multi-level system is obtained, but without using deep characterisation. However, one can apply further flattenings using graph transformation, as shown in Fig. 8:

1. First, one can remove the double linguistic/ontological typing, keeping just one typing. This can be done by adding the linguistic metamodel on top of the ontological stack, and replicating such metamodel elements at all metalevels except 0 and 1.
2. Second, one can flatten a multi-level system into two metalevels. The first variant of this flattening is to merge all models of the ontological stack into a single model, and then consider this merged model as an instance of the linguistic metamodel.

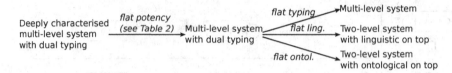

Fig. 8. Flattenings for multi-level systems

The second variant is to merge all models of the ontological stack, except the top-most one, into a single model, and then consider this merged model as an instance of the top-most model. In this variant, given a system like the one in Fig. 1(b), one would obtain a two-level system like the one in Fig. 1(a).

5   Related Work

A first strand of research focuses on multi-level metamodelling. In [12], MOF is extended with multiple metalevels to enable XML-based code generation. Nivel [2] is a double metamodelling framework based on the weighted constraint rule language (WCRL). XMF [7] is a language-driven development framework allowing an arbitrary number of metalevels. Another form of multi-level metamodelling can be achieved through powertypes [13], since instances of powertypes are also subtypes of another type and hence retain both a type and an instance facet. Multi-level metamodelling can also be emulated through stereotypes, although this is not a general modelling technique since it relies on UML to emulate the extension of its metamodel. The interested reader can consult [5] for a thorough comparison of potencies, powertypes and stereotypes.

A second strand of research focuses on deep characterisation. DeepJava [15], META-DEPTH [16], and the works of Gutheil [14], Atkinson [3], and Aschauer [1] all support deep characterisation through potency. While these works agree on that clabjects are instantiated using the multi-potency semantics, they differ in other design decisions. Firstly, some works are ambiguous about the instantiation semantics for associations. In [15], the associations can be represented as Java references; hence we interpret that they are instantiated using the single-potency semantics. In [14], the connectors are explicitly represented as clabjects but their instantiation semantics is not discussed; hence we interpret that they are instantiated using the multi-potency semantics. Secondly, not all works adhere to *strict* ontological metamodelling. In [1], the ontological type of an association does not need to be in the adjacent metalevel above, but several metalevels above. Note that our single-potency semantics makes it possible to retain strict metamodelling for associations through a flattening construction that replicates these associations. Finally, some works differ in how they tackle potency on constraints and methods. Potency on constraints is not explicitly shown in [3] and not considered in [1], whereas potency on methods is only supported by DeepJava and METADEPTH.

Table 3 shows a summary of the particular semantics for deep characterisation implemented by the above mentioned works and compares it with the semantics supported by our formalisation. It is worth noting that no tool recognises the fact that multiplicity constraints are constraints as well and hence can have a potency.

Table 3. Comparison of different deep characterisation semantics

	Clabjects	Associations	Strictness	Constraints	Mult. constraints
DeepJava [15]	▲	△	yes	△	N.A.
Atkinson et al. [3]	▲	▲	yes	▲	▲1
Aschauer et al. [1]	▲	▲	no	N.A.	▲1
METADEPTH [16]	△, ▲	△, ▲	yes	△	▲1
DPF formalisation	△, ▲	△, ▲	yes	△, ▲	△, ▲

6   Conclusion and Future Work

In this paper, we presented a formalisation of concepts of deep metamodelling using DPF and graph transformation. In particular, we provided a precise definition of the double linguistic/ontological typing and two different semantics for potency on different model elements, as well as an operational semantics of potency using a flattening based on graph transformation.

We believe that distinction of two possible semantics for potency is important to achieve more flexible tools, allow their comparison and their interoperability. For instance, in our case study (see Fig. 1), the multi-potency on the association datalink has the effect that one can only add associations between instances of the component Table and instances of the component Map in the model M_3. On the contrary, a single-potency on the association datalink would have the effect that one could add associations between any two component instances in the model M_3, not necessarily between instances of Table and instances of Map. We found both semantics especially useful in this case study.

To the best of our knowledge, this work is the first attempt to clarify and formalise some aspects of the semantics of deep metamodelling. In particular, this work explains different semantic variation points available for deep metamodelling, points out new possible semantics, currently unexplored in practice, as well as classifies the existing tools according to these options.

In the future, we plan to investigate the effects of combining different potency values to interdependent model elements. This includes the investigation of the effects of overriding the potency of a clabject using inheritance, as this may lead to contradictory combinations of potencies. We also plan to formalise the linguistic extensions proposed in [16]. Finally, we will incorporate the lessons learnt from this formalisation into the METADEPTH tool, in particular the possibility of assigning potency to multiplicity constraints.

Acknowledgements. Work partially funded by the Spanish Ministry of Science (project TIN2008-02081), and the R&D programme of the Madrid Region (project S2009 /TIC-1650).

References

1. Aschauer, T., Dauenhauer, G., Pree, W.: Multi-level modeling for industrial automation systems. In: EUROMICRO 2009, pp. 490–496. IEEE Computer Society (2009)

2. Asikainen, T., Männistö, T.: Nivel: A metamodelling language with a formal semantics. Software and Systems Modeling 8(4), 521–549 (2009)
3. Atkinson, C., Gutheil, M., Kennel, B.: A flexible infrastructure for multilevel language engineering. IEEE Transactions on Software Engineering 35(6), 742–755 (2009)
4. Atkinson, C., Kühne, T.: Rearchitecting the UML infrastructure. ACM Transactions on Modeling and Computer Simulation 12(4), 290–321 (2002)
5. Atkinson, C., Kühne, T.: Reducing accidental complexity in domain models. Software and Systems Modeling 7(3), 345–359 (2008)
6. Barr, M., Wells, C.: Category Theory for Computing Science, 2nd edn. Prentice-Hall (1995)
7. Clark, T., Sammut, P., Willans, J.: Applied Metamodelling: A Foundation for Language Driven Development, 2nd edn., Ceteva (2008)
8. Diskin, Z.: Mathematics of Generic Specifications for Model Management I and II. In: Encyclopedia of Database Technologies and Applications, pp. 351–366. Information Science Reference (2005)
9. Diskin, Z., Kadish, B., Piessens, F., Johnson, M.: Universal Arrow Foundations for Visual Modeling. In: Anderson, M., Cheng, P., Haarslev, V. (eds.) Diagrams 2000. LNCS (LNAI), vol. 1889, pp. 345–360. Springer, Heidelberg (2000)
10. Diskin, Z., Wolter, U.: A diagrammatic logic for object-oriented visual modeling. In: Proc. of the 2nd Workshop on Applied and Computational Category Theory (ACCAT 2007). ENTCS, vol. 203(6), pp. 19–41. Elsevier (2008)
11. Ehrig, H., Ehrig, K., Prange, U., Taentzer, G.: Fundamentals of Algebraic Graph Transformation. Springer (March 2006)
12. Gitzel, R., Ott, I., Schader, M.: Ontological extension to the MOF metamodel as a basis for code generation. Computer Journal 50(1), 93–115 (2007)
13. Gonzalez-Perez, C., Henderson-Sellers, B.: A powertype-based metamodelling framework. Software and Systems Modeling 5(1), 72–90 (2006)
14. Gutheil, M., Kennel, B., Atkinson, C.: A Systematic Approach to Connectors in a Multi-level Modeling Environment. In: Czarnecki, K., Ober, I., Bruel, J.-M., Uhl, A., Völter, M. (eds.) MODELS 2008. LNCS, vol. 5301, pp. 843–857. Springer, Heidelberg (2008)
15. Kühne, T., Schreiber, D.: Can programming be liberated from the two-level style? Multi-level programming with DeepJava. In: OOPSLA 2007: 22nd Annual ACM SIGPLAN Conference on Object-Oriented Programming, Systems, Languages and Applications, pp. 229–244. ACM (2007)
16. de Lara, J., Guerra, E.: Deep Meta-modelling with METADEPTH. In: Vitek, J. (ed.) TOOLS 2010. LNCS, vol. 6141, pp. 1–20. Springer, Heidelberg (2010)
17. Rossini, A.: Diagram Predicate Framework Meets Model Versioning and Deep Metamodelling. Ph.D. thesis, Department of Informatics, University of Bergen, Norway (2011)
18. Rossini, A., Rutle, A., Lamo, Y., Wolter, U.: A formalisation of the copy-modify-merge approach to version control in MDE. Journal of Logic and Algebraic Programming 79(7), 636–658 (2010)
19. Rutle, A.: Diagram Predicate Framework: A Formal Approach to MDE. Ph.D. thesis, Department of Informatics, University of Bergen, Norway (2010)
20. Rutle, A., Rossini, A., Lamo, Y., Wolter, U.: A formal approach to the specification and transformation of constraints in MDE. Journal of Logic and Algebraic Programming 81(4), 422–457 (2012)
21. Steinberg, D., Budinsky, F., Paternostro, M., Merks, E.: EMF: Eclipse Modeling Framework 2.0., 2nd edn. Addison-Wesley Professional (2008)

Reusable Graph Transformation Templates

Juan de Lara and Esther Guerra

Department of Computer Science
Universidad Autónoma de Madrid, Spain
{Juan.deLara,Esther.Guerra}@uam.es

Abstract. Model-Driven Engineering promotes models as the principal
artefacts of the development, hence model transformation techniques –
like graph transformation – become key enablers for this development
paradigm. In order to increase the adoption of Model-Driven Engineering in industrial practice, techniques aimed at raising the quality and
productivity in model transformation development are needed.

In this paper we bring elements from *generic programming* into graph
transformation in order to define generic graph transformations that can
be reused in different contexts. In particular, we propose the definition
and instantiation of graph transformation *templates* whose requirements
from generic types are specified through so-called *concepts*, as well as
mixin layers that extend meta-models with the extra auxiliary elements
needed by templates.

1 Introduction

Model-Driven Engineering (MDE) is a software development paradigm that promotes the use of models as the principal assets of the development. Hence, model
manipulation techniques – like graph transformation (GT) [3] – become enabling
technologies for this approach.

In order to foster the use of MDE in industry, techniques aimed at raising the
quality of the generated software and to speed up the productivity of engineers
are needed. One way to improve productivity is to increase the *reusability* of
model transformations, so that they can be applied in different contexts. Unfortunately, building transformations in MDE is a type-centric activity, in the sense
that transformations are defined over the specific types of concrete meta-models,
and it is difficult to reuse them for other meta-models even if they share characteristics. For example, even if many languages share the semantics of Petri nets,
like activity diagrams or process modelling languages, such semantics is nor-
mally defined over some specific meta-model (a concrete realization of a Petri
net meta-model) and cannot be easily reused for other meta-models.

The present work aims at providing mechanisms to enable the *correct* reuse
of GT systems across different meta-models. For this purpose, we build upon
some ideas from generic programming [5] to define generic GT systems that we
call GT *templates*. These generic GT systems are not defined over the types of
concrete meta-models, but over variable types that need to be bound to types

A. Schürr, D. Varró, and G. Varró (Eds.): AGTIVE 2011, LNCS 7233, pp. 35–50, 2012.

of some specific meta-model. However, not every meta-model qualifies as a valid binding for the variable types used in a GT template. Hence, in order to ensure a *correct* reuse, we specify the requirements that meta-models need to satisfy using a so-called *concept* [10,6]. A concept gathers the structural requirements that need to be found in a meta-model to be able to instantiate a GT template on the meta-model types and apply the template to the meta-model instances.

In addition, GTs sometimes need auxiliary model elements to perform some computations. For example, in order to define the semantics of Petri nets, we may need an edge referencing the transition that is currently being fired, or in object-oriented systems we may need to introduce an auxiliary edge to flatten the inheritance hierarchy. These extra elements do not belong to the meta-model of the language, but are auxiliary devices needed by the transformation. Hence, a GT template cannot demand specific meta-models to include such extra devices as part of its requirements. Instead, we define so-called *mixin layers* [10,15]. These are meta-models with parameters, that are *applied* to specific meta-models, increasing them with extra elements by a gluing construction. Mixins are generic, and hence applicable to any meta-model that satisfies the requirements given by a set of concepts. In this way, a GT template can be defined over the types of the mixin and the types of the concepts such mixin needs.

This paper continues our research on model transformation reuse by means of genericity [10,13]. While in [13] we added genericity to model-to-model transformations expressed in the ATL language, here we focus on in-place transformations expressed using GT. The use of a formal framework helps in formulating template instantiations (i.e., bindings) precisely, identifying the needed conditions for correct template reuse, and understanding the composition mechanism of mixins. The formal semantics of DPO graph transformation yields tighter conditions for correct template reuse than in our previous works [10,13].

Paper Organization. First, Section 2 introduces our approach. Then, Section 3 defines meta-models algebraically. Section 4 explains how to bind concepts to meta-models, providing a mechanism to instantiate GT templates. Section 5 shows how to define and apply mixins. Section 6 provides further examples. Finally, Section 7 compares with related work and Section 8 concludes.

2   Overview of the Approach

Frequently, very similar transformations are developed for different meta-models. The reason is that although similar, each transformation is developed to work with the types of a particular meta-model, so that its use with types of other unrelated meta-models is not possible. This results in a waste of effort as the same problems and solutions have to be tackled repeatedly. For example, there is a catalogue of well-known refactorings for object-oriented systems [4]; however, if we encode them as GT rules, we need a different encoding for each meta-model we use. In this way, we need to encode a different transformation for the UML meta-model, the Java meta-model, or the meta-model of any other object-oriented notation we may like to work with.

Analogously, there are languages with similar semantics. For example, many languages share the semantics of Petri nets, such as activity diagrams or domain-specific languages for manufacturing (parts are produced and consumed at machines) and networking (packets are sent and received by computers). However, if we specify their semantics through a GT system, we need to encode a different system for the meta-model of each language.

Therefore, a mechanism to define GT systems for families of meta-models sharing some requirements would promote the reutilization of transformations. For this purpose, we use so-called *concepts* [10,6] to gather the requirements of a family of meta-models, needed by a reusable GT system to work. These requirements are structural, and hence a concept has the form of a meta-model as well. However its elements (nodes, edges, attributes) are interpreted as variables to be *bound* to elements of specific meta-models. The rules of a *GT template* use the variables in the concept instead of the types of a specific meta-model. As a result we obtain reusability because the concept can be bound to a family of meta-models, and the GT template becomes applicable to any of them.

This situation is illustrated in Fig. 1, where a GT template has been defined over the variable types of a concept C. As an example, C may define the core structural elements that characterize Petri net-like languages, and the transformation may include rules (defined over the type variables in C) to refactor Petri net-like models, according to the catalogue in [12]. The concept can be bound to a set $L(C)$ of meta-models sharing the structure required by the concept. In this way, the rules can be applied on instances of any meta-model $MM \in L(C)$. We have depicted the set of models conformant to a meta-model MM as $L(MM)$.

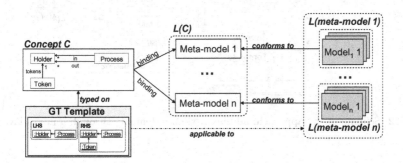

Fig. 1. Scheme of the approach: concept, GT template and binding

In addition, oftentimes, GTs make use of auxiliary graph elements to implement some model manipulations. For example, when defining the semantics of Petri nets, we need an auxiliary node to mark which transition is being fired, as well as the processed input and output places, in order to add/remove tokens to/from the appropriate places. The types of these auxiliary elements do not belong to the meta-model of the language (Petri nets), but are auxiliary elements needed just for the simulation. If this simulator is built as a GT template over a concept, then this concept cannot include the auxiliary elements (and no binding will be provided for them) because the meta-models will hardly ever include such extra devices.

In order to solve this problem, we define so-called *mixin layers* as an extension mechanism for meta-models. A *mixin* is a generic meta-model defining all extra elements needed by a GT template but which are not present in a concept. In addition, it includes some formal parameters acting as gluing points between the mixin and the concept. Thus, the GT template can use the variable types defined on both the mixin and the concept. Once we bind the concept into a specific meta-model, this is extended with the new types defined by the mixin, and the template can be applied on instances of this extended meta-model.

This scheme is shown in Fig. 2. In particular, the mixin adds some auxiliary elements to simulate Petri net-like languages (a pointer to the process being fired and to the processed holders of tokens). The mixin defines as gluing points the nodes **Holder** and **Process**, whose requirements are given by a concept C (structure of Petri net-like languages). Binding the concept to a specific meta-model will increase the meta-model with the elements of the mixin. The GT template is defined over the types resulting from gluing C and the mixin, and is applicable to instances of meta-models to which we can bind C.

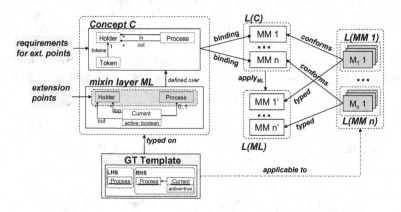

Fig. 2. Scheme of the approach: adding auxiliary elements through a mixin layer

3   An Algebraic Setting for Models and Meta-models

In our setting, we consider attributed,[1] typed graphs [3] of the form $G = \langle V; E; A; D; src_E, tar_E \colon E \to V; src_A \colon A \to V; tar_A \colon A \to D \rangle$, made of a set V of vertices, a set E of edges, a set A of attributes, a set D of data nodes, and functions src and tar that return the source and target vertices of an edge, and the owning vertex and data value of an attribute.

In order to represent meta-models, we consider type graphs with inheritance and with cardinality constraints in associations, in the style of [16]. In this way, a

---

[1] For simplicity, we do not consider abstract nodes or attributes in edges.

meta-model $MM = \langle G; I \subseteq V \times V; card\colon E \to \mathbb{N} \times (\mathbb{N} \cup \{*\}) \rangle$ is made of a graph G, together with a set I of inheritance relations, and a function $card$ that returns the cardinality of target ends of edges. We assume EMF-like references for edges (i.e., with cardinality only in the target end) so that UML-like associations (i.e., with cardinality in both ends) should be modelled as a pair of edges in both directions. Given a node $n \in V$, we define its clan as the set of its direct and indirect children, including itself. Formally, $clan(n) = \{n' \in V | (n', n) \in I^*\}$, where I^* is the reflexive and transitive closure of I. For simplicity, we avoid adding an algebra to MM.

Similar to [16], we give semantics to cardinality constraints by means of positive and negative atomic graph constraints [3] of the form $PC(a\colon P \to Q)$ and $\neg PC(a\colon P \to Q)$. The former require that for each occurrence of P in a graph G, we find a commuting occurrence of Q. Formally, for each $m\colon P \to G$ we need $m'\colon Q \to G$ s.t. $m = m' \circ a$. Negative atomic constraints demand that for each occurrence of P in a graph, there is no commuting occurrence of Q. If G satisfies a constraint a, we write $G \models a$.

In particular, we generate graph constraints regulating the minimum and maximum number of instances at association ends. Thus, for each edge e in the meta-model with $card(e) = [l, h]$, if $l > 0$ we generate a positive graph constraint as shown to the left of Fig. 3, and if $h \neq *$ we generate a negative graph constraint as shown to the right of the same figure.

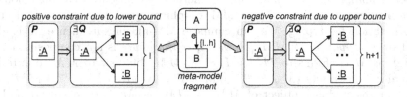

Fig. 3. Graph constraints generated from cardinality constraints

Next we define morphisms between meta-models as a graph morphism with some extra constraints given by the inheritance hierarchy. We will use this notion later to define the binding between a concept and a meta-model.

A morphism $MM \to MM'$ between two meta-models (short MM-morphism) is given by a clan morphism [9] $f\colon G_{MM} \to MM'$ from the graph G_{MM} of the first meta-model to the second meta-model, preserving the inheritance hierarchy. A clan morphism is similar to a standard E-Graph morphism [3], but it also takes into account the semantics of inheritance. Hence, for each edge e of G_{MM} that is mapped to an edge e' of $G_{MM'}$, we allow the source node of e to be mapped to any node in the clan of the source node of e'. Formally, $f_V(src_E(e)) \in clan(src'_E(f_E(e)))$, and similar for the target of edges. In addition, we allow mapping an attribute of a node to an attribute of a supertype the node is mapped into. Formally, $f_A(src_A(a)) \in clan(src'_A(f_A(a)))$. As in [7], the morphism has to preserve the inheritance hierarchy as well, hence if $(u, v) \in I_{MM}$, then $(f(u), f(v)) \in I^*_{MM'}$. Please note that we purposely neglect cardinality constraints in MM-morphisms because the semantics of these is given by graph constraints. We will deal with this issue when defining the binding between a concept and a meta-model.

A model M can be seen as a meta-model with empty inheritance hierarchy and no cardinality constraints. Therefore, we can represent the typing function $M \xrightarrow{type} MM$ as an MM-morphism. In addition, we say that M *conforms to MM* (written $M \models^{type} MM$) if there is a typing $M \xrightarrow{type} MM$ and M satisfies all graph constraints derived from the cardinality constraints in MM.

Example. Fig. 4 shows an MM-morphism f between two meta-models, where we have represented attributes as arrows to a datatype (see **a**) and mapped elements with primas (e.g., node A is mapped to node A'). The attribute in node B is mapped to an attribute defined in E, which is a supertype of the node mapped to B, and the same for the edge e. Regarding the preservation of the inheritance hierarchy, MM-morphisms permit introducing intermediate nodes in the hierarchy of

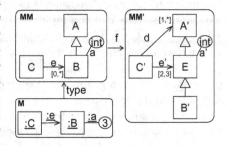

Fig. 4. MM-morphisms

the target meta-model (like node E which is between the image nodes A' and B') as well as mapping several nodes in an inheritance relation into a single node.

The figure also shows a typing MM-morphism $type: M \to MM$ using the UML notation for typing. We can compose MM-morphisms, hence M is also typed by $f \circ type: M \to MM'$. However, *conformance* is not compositional in general, as M conforms to MM ($M \models^{type} MM$) but not to MM' due to its cardinality constraints ($M \not\models^{f \circ type} MM'$). Finally, given $M' \models^{type} MM'$, we have that the pullback object of $MM \to MM' \leftarrow M'$ is typed by MM, but need not be conformant to MM.

4   Graph Transformation Templates

GT templates are standard GT systems specified over the variable types of a concept, which has the form of a meta-model. As an example, Fig. 5 shows to the left the concept *TokenHolder*, which describes the structural requirements that we ask from Petri net-like languages, namely the existence of classes playing the roles of token, holder (places in Petri nets) and process (transitions). We can use this concept to define generic GT systems for the simulation and refactoring of models in languages with this semantics. For instance, the right of the same figure shows one of the behaviour-preserving refactoring rules proposed in [12] expressed in a generic way, using the types of the concept. The rule removes self-loop holders with one token provided they are connected to a single process.

In order to use the GT template, we need to *bind* the concept to a specific meta-model. This binding is an MM-morphism with some extra constraints derived from the particular GT template to be reused. As an example, Fig. 6 shows

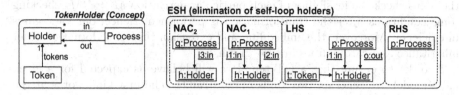

Fig. 5. Concept *TokenHolder* (left), and GT template over the concept (right)

a binding attempt from the *TokenHolder* concept to a meta-model to define factories. *Factory* models contain machines interconnected by conveyors which may carry parts. Both conveyors and machines need to be attended by operators. Hence, our aim is to apply the rule template *ESH* on instances of the *Factories* meta-model. However, there is a problem because this rule deletes holders, and we have bound holders to conveyors, which are always connected to some operator as required by the cardinality constraints. Thus, if we try to apply the rule (using DPO semantics), it will always fail as we will obtain a dangling edge. This shows that the binding should ensure some correctness conditions, which we present in the remaining of the section.

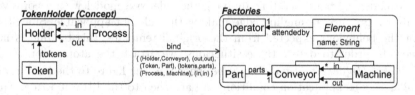

Fig. 6. Binding a concept to a meta-model: first attempt

The general instantiation scheme of a GT template for a given binding is shown in the diagram to the right. The template is typed over the concept C, which needs to be bound to a meta-model MM by a special kind of MM-morphism $bind: C \to MM$. The binding provides a re-typing for the rules of the template, which then become applicable to instance graphs of MM. In order to ensure

$$C \xrightarrow{\;bind\;} MM$$
$$\left(\begin{array}{c} \uparrow \\ type' \end{array} \quad P.B. \quad \begin{array}{c} \uparrow \\ type \end{array}\right)$$
$$M' \text{—} emb \text{➤} M$$
$$\Big\Downarrow* \qquad\qquad \Big\Downarrow*$$
$$M'_f \text{—} emb' \text{➤} M_f$$

a correct application of the template, we must guarantee that for every model $M \models^{type} MM$, if we consider only the elements M' given by the concept C (obtained by the pullback in the diagram), and $M' \models C$, then, for each possible sequence of rule applications over M', there is a sequence of rule applications (using the same order) applicable to M. Moreover, there should be an embedding from the final model M'_f into M_f.

Although this issue resembles the one handled by the embedding and extension theorems [3], there are fundamental differences. The main one is that our goal is to discard invalid bindings and initial models that would lead to an incorrect application of the GT template *before* applying it. On the contrary, the mentioned theorems check the feasibility of each particular derivation $M' \Rightarrow^* M'_f$ checking some conditions *after* the transformation is performed. Hence, our view gains in efficiency. Moreover, the embedding and extension theorems do not consider inheritance and cardinalities in type graphs.

In order to ensure that a GT template will behave as expected for a given meta-model, we generate two kinds of constraints. The first kind works at the meta-model level and forbids bindings that would *always* lead to an incorrect execution of the template, as some rules of the template would be inapplicable for any possible model due to dangling edges or violations of cardinality constraints. However, some bindings may lead to incorrect executions only for *some* initial models. Thus, our second kind of constraints detects potential incorrect template executions for a particular instance of a meta-model. If a binding and an initial model satisfy these constraints, then the template can be safely applied to the model. Next we explain in detail each constraint type.

Constraints for Bindings. These constraints act as application conditions for the *binding*. Fig. 7(a) illustrates how they work. Assume we want to apply our refactoring GT template to models that conform to the *Factory* meta-model. The first step is therefore binding the *TokenHolder* concept to the meta-model, as indicated in the figure. Looking at the rule in Fig. 5 we notice that it deletes holders. As holders are mapped to conveyors, and conveyors are always attended by one operator, applying the rule to a model with a conveyor will always produce a dangling edge making the rule not applicable. Since this is not the behaviour expected from the original template, this binding is not allowed. In order to detect these situations, we attach the atomic constraint $\neg PC(TokenHolder \xrightarrow{q} MandatoryEdge)$ shown in the figure to the binding, in a similar way as application conditions are attached to the LHS of a rule. This constraint forbids a binding if the holder is connected to some node Z through an edge with lower cardinality bigger than 0. As this is the case should we identify e and `attendedby`, the binding is not allowed.

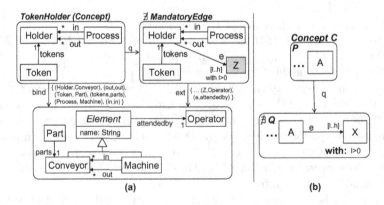

(a)   **(b)**

Fig. 7. (a) Evaluating constraint in a binding. (b) Scheme of constraints for bindings.

The structure of the generated constraints for bindings is shown in Fig. 7(b). The constraints use MM-morphisms and must be satisfied by the function *bind*. They are generated for each node type that is created or deleted by the GT template. Thus, if the template creates or deletes an object of type A, we generate a constraint that has the concept C as premise, and the concept with a class X (existing or not in C) connected to A through an edge e with lower cardinality bigger than 0 as consequence. We also demand that the match of the edge e does not belong to $bind(C)$. If a rule deletes an A object, the intuitive meaning is that it will produce a dangling edge because of this mandatory edge, disabling the rule application. This breaks the correctness criteria because we could apply the rule in a model M' conformant to the concept but not in a model M conformant to the meta-model. Therefore we forbid such a binding. If a rule creates an A object, we generate the same constraint because applying the grammar to a model M typed over MM would only produce incorrect models, as the created object would need to be connected to an object of type X through an unforeseen edge type e. Please note that the generated constraint takes into account the case where the edge e has been defined in an ancestor of the class bound to A, as we use MM-morphisms. On the contrary, the constraint does not detect if there is a subtype defining an unbounded mandatory edge and does not forbid the binding in such a case. Indeed, in this situation, there may be initial models where the template can be safely applied, therefore this scenario is handled by a second set of constraints working at the model level (see below).

Regarding cardinalities, we forbid binding edges with cardinality $[l..h]$ to edges with cardinality $[l'..h']$ if both intervals do not intersect, as from a model $M \models MM$ we would never obtain a model $M' \models C$ performing the pullback. In addition, if a GT template creates or deletes instances of the source or target classes of an edge e defined in a concept C, then we can map the edge to $e' = bind(e)$ only if $card(e) = card(e')$. This condition is not required for edges whose source and target are not created or deleted by the template; however, the initial model must satisfy the cardinality constraints of the concept in any case (i.e., the pullback object M' must satisfy $M' \models C$).

Constraints for Initial Models. These constraints check whether a GT template can be safely executed on a given initial model. For instance, assume that the cardinality of `attendedby` is $[0..1]$ in the meta-model of Fig. 7(a), so that the binding is allowed. Still, given an initial model like the one in Fig. 8(a), our generic refactoring will not be applied to the conveyor as it has one operator, hence leading to a dangling edge as discussed before. This differs from the original template behaviour where such dangling edges do not occur. However, we can safely apply the rule to any model where conveyors have no operator, which we check by generating the constraint in the upper part of the figure.

Fig. 8(b) shows the structure of the generated constraints for initial models. These constraints restrict the instances of the meta-model MM to which the concept C is bound. They are generated for each node type in the meta-model that is deleted by the GT template, as well as for its subtypes, whenever the types declare an edge not included in the binding. Formally, if a generic rule deletes an

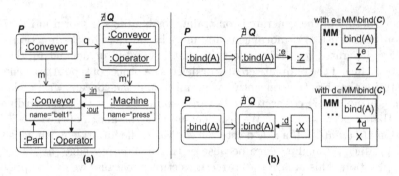

Fig. 8. (a) Evaluating constraint in model. (b) Scheme of constraints for initial models.

object of type B, we generate a constraint for each $A \in clan(bind(B))$ and for each edge e in which A participates whose type belongs to $MM \setminus bind(C)$. This is a sufficient condition to avoid violating the correctness criteria (the instantiated template could fail due to dangling edges) but it is not a necessary condition.

5   Mixin Layers

We now define mixin layers with the purpose of extending meta-models with any auxiliary element needed to execute a GT template. A mixin layer is a meta-model where some of its elements are identified as parameters. Parameters are interpreted as variable types that have to be instantiated to types of the specific meta-model where we want to apply the mixin. However, not every meta-model is eligible to be extended by a particular mixin, and not every type is a valid instantiation of the mixin parameters. The requirements needed by a meta-model and its types are given by one or more concepts.

We define a mixin layer ML as $ML = \langle MM, Conc = \{C_i\}_{i \in I}, Par = \{P_j\}_{j \in J}\rangle$, where MM is a meta-model, $Conc$ is a set of concepts expressing the requirements for meta-models to be extensible by the mixin, and Par is a set of parameters identifying the mixin extension points. Each element P_j in the set of parameters has the form $P_j = \langle G_{MM} \leftarrow G_j \rightarrow G_{C_i}\rangle$, a span relating the graphical elements in the mixin (G_{MM}) with the graphical elements in one of the concepts ($G_{C_i \in Conc}$).

Example. We are building a generic simulator for Petri net-like languages by means of a GT template defined over the concept *TokenHolder*. However, apart from the elements already present in this concept, the simulator must use auxiliary nodes and edges to model the firing of transitions. Adding these elements to the concept is not an option because it would imply that the definition of every language to be simulated with the template should be modified manually to include these auxiliary elements in its meta-model. Instead, we define a mixin which increases any meta-model to be simulated with these auxiliary elements in a non-intrusive way. Fig. 9 shows the mixin (dotted box named "Simulation mixin") which declares two parameters (shaded classes `Holder` and `Process`).

Additionally, the concept *TokenHolder* gathers the requirements for the eligible meta-models for the mixin. The relation between the mixin and the concept is expressed as a span of MM-morphisms. This is used to build the meta-model shown to the right by a gluing construction (a colimit, even though in the particular case of the figure it is also a pushout). This meta-model contains all variable types that the GT template can use.

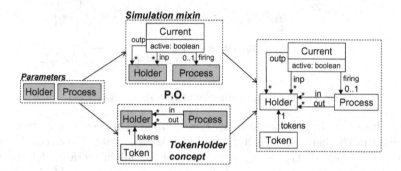

Fig. 9. Specifying a mixin for the simulation of Petri net-like languages

Fig. 10 shows some rules of the generic simulator defined over the mixin. The rule to the left selects one enabled process to be fired, marking it with an instance of the **Current** class from the mixin. Its application condition checks the enabledness of the process (i.e., each input holder has a token). The rule to the right removes one token from an input holder of the process being fired, marking it as processed. Additional rules produce tokens in output holders, and unmark the processed holders and process to allow further firings.

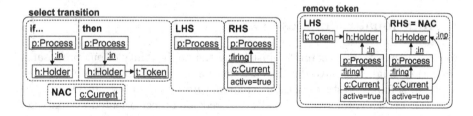

Fig. 10. Some rules of the GT template for the simulation of Petri net-like languages. The template is defined over the mixin *Simulation* shown in Fig. 9.

A mixin becomes applicable by binding its concepts to a meta-model. Fig. 11 shows the binding of concept *TokenHolder* to the *Factory* meta-model. Then, the bound meta-model is extended with the elements defined in the mixin meta-model but not in the concept (class **Current** and edges **inp**, **outp** and **firing**). Thus, we can apply the GT template to instances of the resulting meta-model.

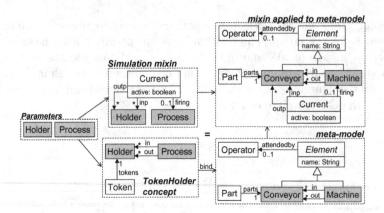

Fig. 11. Applying mixin *Simulation* to a meta-model

The left of Fig. 12 presents formally how a mixin is applied to a meta-model. The mixin in the figure defines a meta-model MM_{ml}, a set of concepts C_i and parameters $MM_{ml} \leftarrow G_j \rightarrow C_i$. The gluing of the mixin meta-model and the concepts is obtained by calculating their colimit,[2] yielding object $\overline{MM_{ml}}$. This is the meta-model over which a GT template is defined. For instance, the template rules shown in Fig. 10 use the meta-model to the right of Fig. 9. Next, the mixin can be applied by binding its concepts to a particular meta-model (or in general to a set of meta-models, as it is not necessary to bind all concepts to the same meta-model). The colimit of the different mixin parameters $MM_{ml} \leftarrow G_j \rightarrow C_i$ and the bound meta-models yields $\overline{MM}$, which is used to retype the template for its application on the bound meta-models. By the colimit universal property, there is a unique commuting $u \colon \overline{MM_{ml}} \rightarrow \overline{MM}$, which acts as binding between the meta-model over which the GT template is specified and the extended specific meta-model. The right of Fig. 12 shows this unique binding for the example.

6   Additional Examples

Next we show a further example illustrating the applicability of our proposal.

Software Engineers have developed a catalog of refactorings to improve the quality of software systems without changing its functionality [4]. One of the most well known catalogs is specially tailored for object-oriented systems [4]. This catalog describes rules applicable to any object-oriented language, but their encoding for a particular object-oriented notation cannot be reused to refactor other notations. Thus, a developer should encode different rule sets for the Java meta-model, the UML meta-model, and so on.

In our approach, we can define the refactorings once over a concept and then bind the concept to several meta-models, obtaining reuse for each bound meta-model. Fig. 13 shows (a simplification of) the concept, together with bindings for two meta-models. The one to the left defines simple UML class diagrams.

---

[2] In the category of MM-objects and MM-morphisms.

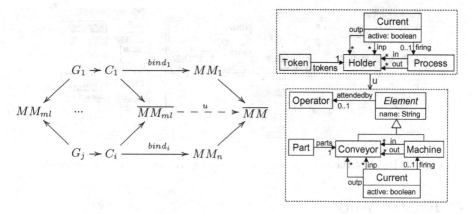

Fig. 12. Binding and applying a mixin (left). Resulting binding for the example (right).

The right meta-model is for Rule-Based Access Control (RBAC) [14], and permits the definition of properties and permissions for roles that can be hierarchically arranged. Children roles inherit the properties and permissions of parent roles.

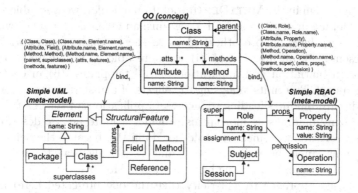

Fig. 13. Concept for object-oriented systems, and two possible bindings

Fig. 14 shows the generic rules for the *pull-up attribute* refactoring [4], which moves an attribute to a superclass when all its children classes define it. The first rule detects the refactoring opportunity and moves the attribute from one of the children classes to the superclass. Then, the second rule removes the attribute from the rest of children. We can also implement a more refined version of this refactoring by defining a mixin that declares a pointer for classes, which the rules can use to indicate the parent class being refactored (i.e., class p in the rules).

The binding $bind_1$ in Fig. 13 allows applying the generic refactoring to UML models and pull-up fields. Nonetheless, using a different binding permits refactoring references and methods as well, mapping `Attribute` to `Method` or to `Reference`. Similarly, the binding $bind_2$ permits refactoring properties in role hierarchies, but we can bind `Attribute` to `Operation` to pull-up operations.

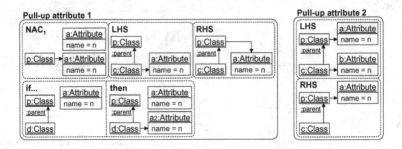

Fig. 14. Rules of the GT template implementing refactoring *pull-up attribute*

7   Related Work

In object-oriented programming, mixins and traits are classes that provide extra functionality without being instantiated. Instead, other classes may inherit from the mixin, which is a means to collect functionality. We have generalized this idea to mixin layers. These are parameterised meta-models adding auxiliary elements to a set of meta-models sharing the characteristics specified by a concept.

In previous work [10], we brought ideas from generic programming into MDE, implementing them in the METADEPTH tool. In particular, we were able to write generic model manipulations as EOL programs, an OCL-like language with side effects. Here we have adapted these ideas to the algebraic framework of GT, which presents several advantages: (i) we were able to formulate correctness criteria for instantiation and application of GT templates; (ii) in contrast to EOL programs, GT permits analysing the effects of transformations, and this is useful to discard bindings leading to incorrect applications of the GT templates; (iii) formalizing bindings as morphisms provides a more precise description of the binding constraints, which we could not do in [10] because the behaviour of EOL programs cannot be easily analysed; and (iv) the algebraic formalization of mixins helped us in understanding how they work. Moreover, we discovered that a pattern-based approach to genericity (like the one presented here) imposes less restrictive conditions for the binding than one based on a scripting language (like EOL or OCL). This is so as MM-morphisms allow defining the target of a reference in a supertype (cf. reference e' in Fig. 4). In EOL, a navigation expression $c.e$ may lead to an E object, which may not have all properties of a B' object (as expected by the generic operation). On the contrary, in GT one provides a pattern with an explicit type for its objects (e.g., B which gets mapped to B'), hence filtering the undesired E objects.

Parameterized modules were proposed in algebraic specification in the eighties [2]. A parameterized module is usually represented with a morphism $par: P \to M$ from the formal parameters to the module. In this paper, we propose using *concepts* to restrict how the formal parameters can be bound to the actual parameters in mixins. We can also think as GT templates as parameterized models (by a concept). In this case, the special semantics of DPO GT induce additional constraints in the binding. Our MM-morphisms are based on S-morphisms [7], but we support attributes and do not require morphisms to be subtype preserving. Our composition mechanism is also related to Aspect-Oriented Modelling [8], which focuses on modularizing and composing crosscutting concerns.

In the context of GT, there are some proposals for adding genericity to rules. For example, the VIATRA2 framework [1] supports generic rules where types can be rule parameters. MOFLON has also been extended with generic and reflective rules using the Java Metadata Interface [11]. These rules can receive string parameters that can be composed to form attribute or class names, and may contain nodes that match instances of any class. Still, none of these tools provide mechanisms (like *concepts* and *bindings*) to control the correctness of rule applications, or extension mechanism (like mixins) for meta-models. We believe that the ideas presented in this paper can be adapted to these two approaches.

8   Conclusions and Future Work

In this paper we have adapted *generic programming* techniques to increase the reusability of GTs. In particular, we have defined GT templates, which are not typed over a specific meta-model, but over *concepts* specifying the structural requirements that a meta-model should fulfil if we want to apply the template on its instances. Hence, the GT template can be instantiated for any meta-model satisfying the concept, obtaining reusable transformations. Besides, many GT systems use auxiliary elements that have to be included ad-hoc in the meta-models. We have proposed a non-intrusive solution consisting on the definition of mixin layers declaring any extra device used by the template. The requirements that a meta-model should fulfil to be extended through the mixin are given by a concept. Again, we obtain reusability because a template that uses types of a mixin can be applied to any meta-model that satisfies the mixin requirements.

As for future work, currently we forbid bindings that can lead to an incorrect execution of a GT template. However, it may be sometimes possible to semi-automatically adapt the template to make it work correctly. In addition, now we require an exact match of the concept in the meta-models, but to increase reusability, we plan to provide techniques to resolve some heterogeneities in the binding, in the line of [13]. On the practical side, we plan to include the lessons learnt regarding correct binding and correct reuse in our METADEPTH tool.

Acknowledgements. Work funded by the Spanish Ministry of Economy and Competitivity (project TIN2011-24139) and the Region of Madrid (project S2009 /TIC-1650).

References

1. Balogh, A., Varró, D.: Advanced model transformation language constructs in the VIATRA2 framework. In: Proc. of the 21st Annual ACM Symposium on Applied Computing, SAC 2006, pp. 1280–1287. ACM (2006)

2. Ehrig, H., Mahr, B.: Fundamentals of Algebraic Specification 2: Module Specifications and Constraints. Monographs in Theoretical Computer Science. An EATCS Series, vol. 21, Springer (1990)
3. Ehrig, H., Ehrig, K., Prange, U., Taentzer, G.: Fundamentals of Algebraic Graph Transformation. Springer (2006)
4. Fowler, M.: Refactoring: Improving the Design of Existing Code. Addison-Wesley, Boston (1999)
5. García, R., Järvi, J., Lumsdaine, A., Siek, J., Willcock, J.: A comparative study of language support for generic programming. ACM SIGPLAN Notices 38(11), 115–134 (2003)
6. Gregor, D., Järvi, J., Siek, J., Stroustrup, B., Dos Reis, G., Lumsdaine, A.: Concepts: Linguistic support for generic programming in C++. ACM SIGPLAN Notices 41(10), 291–310 (2006)
7. Hermann, F., Ehrig, H., Ermel, C.: Transformation of Type Graphs with Inheritance for Ensuring Security in E-Government Networks. In: Chechik, M., Wirsing, M. (eds.) FASE 2009. LNCS, vol. 5503, pp. 325–339. Springer, Heidelberg (2009)
8. Kienzle, J., Al Abed, W., Fleurey, F., Jézéquel, J.-M., Klein, J.: Aspect-Oriented Design with Reusable Aspect Models. In: Katz, S., Mezini, M., Kienzle, J. (eds.) Transactions on AOSD VII. LNCS, vol. 6210, pp. 272–320. Springer, Heidelberg (2010)
9. de Lara, J., Bardohl, R., Ehrig, H., Ehrig, K., Prange, U., Taentzer, G.: Attributed graph transformation with node type inheritance. Theoretical Computer Science 376(3), 139–163 (2007)
10. de Lara, J., Guerra, E.: Generic Meta-modelling with Concepts, Templates and Mixin Layers. In: Petriu, D.C., Rouquette, N., Haugen, Ø. (eds.) MODELS 2010, Part I. LNCS, vol. 6394, pp. 16–30. Springer, Heidelberg (2010)
11. Legros, E., Amelunxen, C., Klar, F., Schürr, A.: Generic and reflective graph transformations for checking and enforcement of modeling guidelines. Journal of Visual Languages and Computing 20(4), 252–268 (2009)
12. Murata, T.: Petri nets: Properties, analysis and applications. Proceedings of the IEEE 77(4), 541–580 (1989)
13. Sánchez Cuadrado, J., Guerra, E., de Lara, J.: Generic Model Transformations: Write Once, Reuse Everywhere. In: Cabot, J., Visser, E. (eds.) ICMT 2011. LNCS, vol. 6707, pp. 62–77. Springer, Heidelberg (2011)
14. Sandhu, R.S., Coyne, E.J., Feinstein, H.L., Youman, C.E.: Role-based access control models. IEEE Computer 29(2), 38–47 (1996)
15. Smaragdakis, Y., Batory, D.: Mixin layers: An object-oriented implementation technique for refinements and collaboration-based designs. ACM Transactions on Software Engineering and Methodology 11(2), 215–255 (2002)
16. Taentzer, G., Rensink, A.: Ensuring Structural Constraints in Graph-Based Models with Type Inheritance. In: Cerioli, M. (ed.) FASE 2005. LNCS, vol. 3442, pp. 64–79. Springer, Heidelberg (2005)

Towards an Automated 3D Reconstruction
of Plant Architecture

Florian Schöler and Volker Steinhage

Institute of Computer Science III, University of Bonn
Römerstraße 164, 53117 Bonn, Germany
{schoele,steinhag}@iai.uni-bonn.de
http://ivs.informatik.uni-bonn.de/

Abstract. Non-destructive and quantitative analysis and screening of plant phenotypes throughout plants' lifecycles is essential to enable greater efficiency in crop breeding and to optimize decision making in crop management.

In this contribution we propose graph grammars within a sensor-based system approach to the automated 3D reconstruction and semantic annotation of plant architectures. The plant architectures in turn will serve for reliable plant phenotyping. More specifically, we propose to employ Relational Growth Grammars to derive semantically annotated 3D reconstruction hypotheses of plant architectures from 3D sensor data, i.e., laser range measurements. Furthermore, we suggest deriving optimal reconstruction hypotheses by embedding the graph grammar-based data interpretation within a sophisticated probabilistic optimization framework, namely a Reversible Jump Markov Chain Monte Carlo sampling.

This paper presents the design of the overall system framework with the graph grammar-based data interpretation as the central component. Furthermore, we present first system improvements and experimental results achieved in the application domain of grapevine breeding.

Keywords: 3D Reconstruction, Relational Growth Grammar, Reversible Jump Markov Chain Monte Carlo, Plant Phenotyping, Grapevine Breeding.

1 Introduction

This work is part of *CROP.SENSe.net* [4], an interdisciplinary research network whose goal is the assessment of traits of plants for more efficient plant breeding and crop management. The network is subdivided into several projects, each of which is concerned with different approaches. Here we are focussed on the 3D reconstruction of plants and the semantics, i.e., the botanical annotation of all reconstructed plant components from sensor data as a basis for phenotyping.

The foremost aim is a reconstruction algorithm that can handle arbitrary plants. We enhance the strict geometric reconstruction by semantic annotations. These annotations are attached to the components of the plant on the one hand and to the relations between components on the other. They comprise additional information about the component or relation. This information in turn is used to assess phenotypic traits of the plant, like its (estimated) yield. To achieve this we currently work on grapevine as an exemplar

A. Schürr, D. Varró, and G. Varró (Eds.): AGTIVE 2011, LNCS 7233, pp. 51–64, 2012.

plant. There we face several challenges. (1) *Diversity in cultivar architectures:* There is a huge number of different grapevine cultivars, each of which has its own characteristics regarding the architecture like the number of bunches of grapes or the way the trunk grows (straight up or protruding). (2) *Diversity in plant part architectures:* Every grapevine has three parts. The trunk, the leaves and the bunches of grapes. Every such part has a different architecture in different levels of complexity; the leaves have different venations and contours depending on the cultivar, or the bunches of grapes can be more or less compact. (3) *Occlusions:* A reconstruction of the complete plant is very difficult since many parts of it are occluded. When in full bloom one side of the plant only shows the foliage, blocking sight from that perspective. In bunches of grapes one can only sense the outer berries, the inner ones are occluded. (4) *Inconsistent naming:* Because grapevines are grown all over the world sometimes there are different names for the same cultivar or one name is used for different cultivars [21]. For a first approach we concentrate on so-called stem skeletons, i.e., bunches of grapes where the berries were removed. An example can be seen in Fig. 1.

Fig. 1. A sample stem skeleton

The core piece of our approach is a model of the plant architecture for a pruning of the space of all reconstruction hypotheses and integrating this model into a sampling method to intelligently explore the hypothesis space.

For the modeling of plant architectures there are different possibilities. The most popular formalism for the modeling of plants are Lindenmayer systems (L-systems) [10,11]. An L-system is a parallel string rewriting grammar, where one has an alphabet of characters, an arbitrary start word (axiom) over characters from the alphabet and a set of production rules or productions that replace a character by a string. L-systems were first introduced in 1968 as a means to model the development of filamentous organisms. What distinguishes an L-system from other formal grammars is that the production rules are executed in parallel. This is motivated by the plant domain, since all parts of a plant also grow in parallel. The basic L-system formalism suffers from several shortcomings. For example, local context is restricted to a fixed length, the number of relations between components is restricted, there can only be relations between components represented by neighboring characters in the string, or the fact that a one-dimensional string is no well suited representation of three-dimensional, branched plant structures. Some of the shortcomings were eliminated by extensions, like parameterized L-systems, stochastic L-systems or table L-systems [16]. But disadvantages like the limited number of relations or the unsuited representation as a string remain since they are inherent in L-systems.

Therefore we opt for a grammar based functional-structural plant modeling, namely Relational Growth Grammars (RGGs) [8,9], that meets most of the disadvantages of L-systems. A Relational Growth Grammar is a parallel graph transformation system, meaning that, as in an L-system the productions are applied in parallel, only they are applied to graphs rather than strings. This has several benefits for plant modeling in general and for our purposes in particular. Arbitrary relations between arbitrary plant parts can be modeled as labeled edges of the graph, whereas the nodes may be arbitrary components of plants. Furthermore, whole sets of subgraphs can be rewritten with a single production rule, as opposed to replacing a single character with a string. Other advantages are that branches are directly represented, that the context of a rule is not limited in size or location and that therefore local as well as global sensitivity can be modeled. Since strings can be seen as a graph with a linear sequence of nodes (one node for each character in the string), RGGs can be interpreted as a superset of L-systems, meaning that with the RGG formalism one can mimic every L-system, but has even more descriptive power.

With this formalism we can directly include the semantic annotations as well as the relations within our model. For example, a frustum (part of a stem skeleton) can be connected to a sphere (berry) via a *attached_to* relation, or it can be connected to a sphere that forms a joint via a *connected_to* relation. The relations can also be used for the creation of a cohesive volume of the generated structure, see Sect. 4.3.

The remainder of this article is organized as follows. In Sect. 2 we introduce an overview of related work regarding the modeling and reconstruction (of plants) and research on grapevines. Section 3 gives an introductory overview of the approach followed by Sect. 4, which gives a summary of what is already achieved and what is currently being implemented. The article is closed with Sect. 5, that concludes and shows what we are planning for the future of the project.

2   Related Work

Besides L-Systems and RGGs there are the so-called Shape Grammars [20]. Similar to L-systems and RGGs, Shape Grammars have an initial shape and a set of replacement rules that operate on parts of a given shape and replace such a part with a new shape. A difference is that the rule execution does not happen in parallel. On the basis of Shape Grammars Müller et al. [14] developed CGA shape, a Shape Grammar for the procedural modeling of building shells. Ganster and Klein [5] developed an integrated environment and a visual language for the procedural modeling with the aim to ease the development of models.

Regarding the sensing of objects, Schnabel et al. [17] consider cases where objects can only be sensed in a way that the sensor data contain holes in the objects. At first the input point cloud is approximated by primitive shapes (planes, spheres, cylinders, cones, tori). Then, these primitives are used as a guidance to complete unsensed areas and form a complete structure. This approach is limited to situations, where primitives have been detected in the vicinity of holes, whereas our approach is more general and is able to approximately fill parts that are not represented in the sensor data by help of the domain knowledge. For example, in a bunch of grapes we can argue that, depending on the cultivar, the number of reconstructed berries and the sensed size of the bunch, we can expect n more berries that are occluded by the outer berries and we could even approximate their location.

There are other approaches that especially concentrate on the 3D reconstruction of plants. For example, in [1] a method is introduced which also uses some kind of domain knowledge of trees to guide the reconstruction process within three-dimensional laser range data. Their aim is to provide information for the creation of computer models of tree function. In this approach, a single branch consists of a set of fixed-length segments (cylinders) and two angles to its parent. The angles and the radius of a segment are drawn according to certain probability distributions. Also, they use data of structurally simple trees. We do not use fixed length but rather variable length segments depending on the sensor data. Additionally, we are currently gathering knowledge about grapevine cultivars in order to model the plant's architecture more precisely. Moreover, an RGG is capable of modeling the functional as well as the structural dimensions of a plant at the same time and within the same model, rendering the separate functional model unneeded.

In [19] L-systems are used as one step in the reconstruction of foliaged trees based on image data. They segment each image of one tree into tree and background, and construct a skeleton of the tree with only a few branching levels. This skeleton is used as the axiom of an L-system which is then used to generatively add higher order branchings and leaves. They do not leverage the fact that such a grammar can be used as a guidance for the reconstruction.

There is also current research on the modeling and reconstruction of grapevines following different approaches. Keightley and Bawden introduced a method for the volumetric modeling of grapevine biomass [7]. On the basis of three-dimensional laser scan point clouds they estimate the volume of grapevine trunks and compare their estimates with the values gathered by submerging the trunks in water and measuring the overflowing water. The trunks were placed on a turntable and scanned at ten specific angles from a distance of 8.5 m. They concentrate on the scanning itself and how different scans can be combined. In our approach the biomass is only one of several phenotypical features to be determined and therefore only a part of the overall goal.

In [15] a stochastic growth model of grapevine is introduced concentrating on how the plant develops in dependance of the trophic relationships between plant parts and water depletion. Louarn et al. [13] model the canopy structure of grapevines and use this in [12] to analyze the influence of trellis systems and shoot positioning on the canopy structure and light interception. A former investigation of the development of the architecture of inflorescences (early state of a bunch) and bunches was performed by [18]. They analyzed four grapevine cultivars for phenotypical features that contribute to differences in bunch compactness.

3   Overview of the Approach

3.1   General Overview

In the fields of data interpretation and 3D reconstruction the goal is to find a hypothesis that in some way best explains the given data. Usually there is a huge, or even infinite number of possible hypotheses. Therefore one has to find a way to generate hypotheses, prune the space of all hypotheses, and select hypotheses in a smart way.

For the generation of hypotheses and the pruning of the hypthesis space we use the aforementioned Relational Growth Grammar formalism. Hypothesis generation occurs through execution of the production rules. The pruning takes effect through the fact that the productions determine where components are allowed, thus incorporating geometric constraints. Additionally, they also give semantic constraints, e.g., a berry is only allowed at the end of a twig, not in between. Both kinds of constraints are directly encoded into the productions.

We use Reversible Jump Markov Chain Monte Carlo (RJMCMC) [6] sampling for the selection of hypotheses. With this approach we can, within one cohesive framework, select a proper model to interpret the data as well as optimize the parameters of the selected model.

Figure 2 shows an overview of the components of the reconstruction algorithm and a resulting structure. Interpretation of the sensor data is done by optimizing a model-based interpretation that relies on the rules of the RGG. Optimization is done by the RJMCMC-sampling of the hypothesis space. The sampler selects the best generated result. For the sake of clarity, the semantic annotations are not shown in the image.

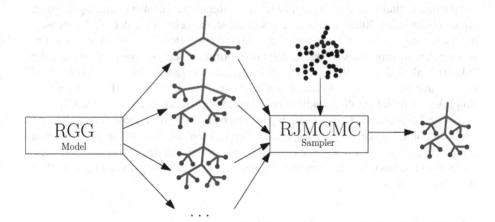

Fig. 2. Components of the reconstruction algorithm and its result. The RGG model generates several reconstruction hypotheses according to the replacement rules. The sampler evaluates them with regards to the sensor data and chooses the best one. The annotations are not shown for the sake of clarity.

3.2   Sensor Data

As sensor data we use three-dimensional point clouds. The point clouds are generated by an ultrahigh precision laser rangefinder Perceptron ScanWorks V5 mounted onto a Romer Infinite 2.0 articulated arm. The rangefinder produces line scans of 7640 points per line at a frequency of 60 Hz and an average point to point resolution of 0.0137 mm. By moving the scanner around the software provided by the vendor accumulates the line scans into one single coordinate frame according to the joint angles of the arm. The result is a three-dimensional point cloud. Figure 3 shows three sample point clouds of the stem skeletons of different cultivars.

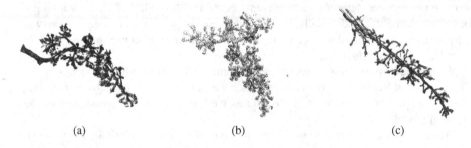

(a)   (b)   (c)

Fig. 3. Sample point clouds of three stem skeletons of different cultivars

3.3   Domain Knowledge

For the interpretation of the sensor data we use a Relational Growth Grammar based on domain knowledge. Rules and parameters of the stem skeleton are derived by measuring the topological skeletons of the corresponding point clouds of the training set. For an overview of properties of skeletonization algorithms see [3]. We currently use the "SkelTre" algorithm by Bucksch [2]. This algorithm organizes the given point cloud by an octree and constructs an initial graph structure over that octree. It then iteratively simplifies the initial graph by finding special node configurations within it, the so-called E-pairs and V-pairs, until no more such pairs can be found. An example of a point cloud and its topological skeleton can be seen in Fig. 4. From the final topological skeleton we then extract features like the lengths of internodes, the angles of internodes to their preceeding internode, or the branching depth of a node. All the gathered data are then stored in a database.

3.4   Modeling with Relational Growth Grammars

As a software tool for the implementation of RGGs we use GroIMP, which was developed alongside the RGG formalism and is available as an open source distribution. It is furthermore easily extensible through a simple plugin interface. GroIMP offers a user-adjustable interface (see Fig. 5 for a screenshot), a text editor with syntax highlighting and different modes for visualization like a wireframe visualization, raytracer based and hardware accelerated OpenGL visualization. GroIMP can read and/or write

Fig. 4. A point cloud and its topological skeleton

a wide range of file formats. It allows direct access to the data structure enabling the user to inspect current values of parameters of nodes in the graph or even to delete parts of the graph. For a plant modeler this is useful, for example, for an analysis of the plant's behavior in the next development steps. Furthermore, it introduces XL (eXtended L-systems) as the programming language for implementing RGGs, which is an extension of the well known Java programming language. Therefore nodes in the graph are objects in the sense of Object Oriented Programming.

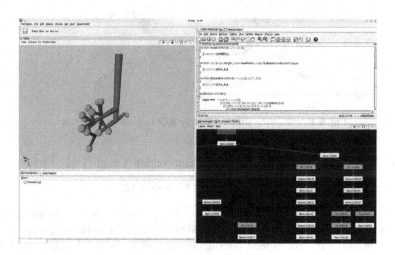

Fig. 5. A screenshot of GroIMP. The interface consists of different, user-adjustable parts. Here one can see a text editor for the grammar code in the upper right. The lower right part shows a part of the current graph. The upper left shows a 3D visualization of the nodes in the graph that correspond to drawable components. The lower left shows a list of files contained in the project.

In general, the structure of an RGG production looks as follows:

$$(*C*), L, (E) ==> R \{P\};$$

C is the context of the rule and L is to be replaced by R. All of those are sets of graphs. E is a set of logical expressions as a condition for the execution of the rule and P is procedural code. An example RGG (taken from GroIMP) implemented in XL looks as follows:

```
protected void init() [
    Axiom ==> A(1);
]

public void run() [
    A(x) ==> F(x)[RU(30) RH(90) A(x*0.8)]
                 [RU(-30) RH(90) A(x*0.8)];
]
```

Where the *init()* method implements the initial structure and the *run()* method a developmental production. *A* is a green sphere of fixed size, *F* is a cylinder of fixed radius and variable length, and *RH* and *RU* are rotations about the local z- and y-axis, respectively. The bracket characters [and] introduce branches in the graph. Figure 6 shows the result after five executions of the *run()* method.

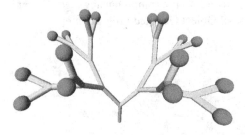

Fig. 6. Result of the simple example RGG after five iterations

A rule with context and an additional Boolean condition might look as follows:

```
(* x:Node *), B, (x instanceof A) ==> F(5) B;
```

This rule says that a node of type *B* is only to be replaced by an *F* node and a new *B* node, if somewhere in the graph there is a node of type *A*. The rule is applicable to the left graph in Fig. 7 but not to the right one. In contrast to the fixed length contexts of L-systems, this RGG rule is applicable no matter how many nodes there are on the path from *B* to the *A* node and they do not even have to be on the same path.

Besides standard replacing rules, XL allows a second kind of rule: execution rules. Those are identified by a ::> arrow and are often used to alter the values of parameters of nodes without replacing them, e.g.,

```
f:Factor <-encodes- g:Gene(gc) ::>
  f[concentration] :+=
  Math.max(0,
          sum(((*Factor(ca,)-Activates(s,m)->g*),
              m * ca / (s + ca))) + gc);
```

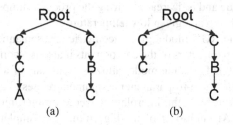

Fig. 7. The production rule *(* x:Node *), B, (x instanceof A)* ==> *F(5) B;* is only applicable to the left graph, not to the right one

This rule (taken from a GroIMP example) searches for all pairs of nodes of types *Factor* and *Gene*, where there is an edge with label *encodes* from the *Gene*-Node to the *Factor*-Node and assigns identifiers *f* and *g*, respectively. It then updates the *concentration* parameter of Node *f* by calling the standard *max* method of the Java Programming Language. Execution rules are useful when only the functional part of a plant (like photosynthesis or the flow of nutrients) has to be updated but not its structure.

This example also demonstrates how arbitrary relations can be used. In the example a gene regulatory network is constructed and the *Activates(,)* relation, which is a Java Class, is used to determine how much a *Gene* is activated or repressed by a *Factor*.

3.5 Reconstruction

As stated, the main challenge is that in reconstruction we have a possibly inifinite number of possible hypotheses to interpret the sensor data. Our two main ingredients for solving the problem are the Reversible Jump Markov Chain Monte Carlo (RJMCMC) sampler on the one hand and the RGG model on the other.

The basic Markov Chain Monte Carlo (MCMC) sampler works by performing so-called jumps within a space of hypotheses, where each hypothesis has the same size, i.e., number of parameters. A jump generates a new hypothesis on the basis of the current one. MCMC therefore samples from the probability distribution over all possible hypotheses by simulating a Markov Chain. But this is not powerful enough for our purposes, since the number of parameters of different hypotheses can vary largely. For such cases Green developed the RJMCMC sampler [6]. This method allows jumps between different hypothesis spaces, where the hypotheses have the same size within one space but differ in size between spaces. In other words, the sampler jumps within the union of all the spaces. An important property of the jumps is that they have to be reversible, i.e., it has to be possible to jump back to earlier states from any given later state. Once a hypothesis is generated it has to be evaluated and decided if it is put into the chain or if it is rejected. If it is accepted the hypothesis is the basis for the hypothesis in the next iteration. Otherwise it is discarded and the next hypothesis is constructed on the basis of the old one.

To make sure, that the reconstruction terminates, i.e., the sampler stops jumping in the hypothesis space, we incorporate Simulated Annealing. There one has a temperature that starts at a high value and is decreased during the process. Jumps are chosen relative to the temperature and stop at a given low temperature

As stated before, the RGG model gives geometric and semantic constraints on the structure. Through the parameters of the components it also is a compact representation of all possible (at least as far as the model allows) exemplars of a certain cultivar. On the other hand a specific value-to-parameter assignment relates to a certain exemplar.

Applied to our task at hand the interplay between sampler, model, and data turns out to be as follows. At iteration t of the algorithm, the sampler randomly chooses a jump that is to be performed next, dependent on the current temperature. Each jump corresponds to one replacement rule in the RGG model and that rule is then executed on the hypothesis h_{t-1}, resulting in a new hypothesis h_t. Now, h_t is evaluated with respect to the data. If it is accepted, h_t is the basis for the next iteration, otherwise h_t is set to h_{t-1}. After an update of the temperature this process is repeated until the temperature falls below a threshold value.

We propose a list of jump/RGG-rule pairs as follows, where frustums are used to model the twigs of the stem skeleton (see Sect. 4.2).

1. Parameter update: Simply changes a randomly chosen parameter of a randomly chosen component.
2. Add branch: Adds a completely new branch to the object. The model defines where and how branches may be added.
3. Delete branch: Deletes one already existing branch.
4. Edit branch: Changes the number of twigs at a branch by adding or deleting twigs.
5. Split frustum: Splits one frustum into two and inserts a sphere in between, to accomodate to the curvature of the object under consideration.
6. Merge frustums: Merges two frustums that are connected via a sphere to a single frustum.

Figure 8 shows a sketch of the reconstruction process. It starts with an initial structure at the initial temperature T_0. At first the topology is changed by adding / deleting branches, resulting in the second depicted hypothesis. In the following further topology changes and parameter updates occur resulting in the last depcited hypothesis at the final temperature T_{final}.

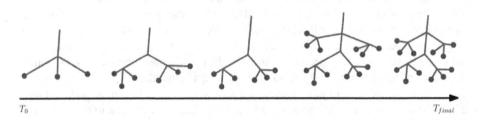

T_0 T_{final}

Fig. 8. Sketch of the reconstruction process: At first mostly topology changes occur, at lower temperatures mostly parameter updates are carried out

4   First Results and Current Developments

4.1   Acquired Data

For a first data acquisition we laser scanned a total of 25 stem skeletons from five differ-ent cultivars. This is already sufficient for evaluating a first approach to reconstruction. We first rely on sensor data gathered in controlled situations. Later we will switch to "real" data gathered in the field.

Additional to the laser scanning we also perform the skeletonization measurements described in Sect. 3.3 and collect all those information in a database.

The reason why we chose stem skeletons for a first data acquisition is that if we want to reconstruct the architecture of a bunch of grapes, stem skeletons are exactly this. Besides, in order to find the berries in the sensor data it is useful to have knowledge about where to expect them. This is exactly what a stem skeleton can tell us.

4.2   Modeling Stem Skeletons

The most basic approach to modeling stem skeletons is to use cylinders for the twigs or parts of twigs. For long twigs one could use a chain of cylinders with decreasing ra-dius. An example of this approach can be seen in Fig. 9(a). This rather simple example already shows the shortcomings of that approach. There are sharp crossovers as well as gaps between cylinders. To generate a more realistic look one could constrain the length of a cylinder, resulting in a greater number of cylinders and therefore an increased com-putation time since more cylinders have to be handled. We therefore follow a different approach. This uses frustums instead of cylinders to cope with the sharp crossovers on the one hand and fills gaps with additional spheres on the other. Figure 9(b) shows the same example as before but with the more advanced modeling approach. As can be seen this generates a much more realistic look. But not only the appearance is more realistic, this model also is a more precise structural representation of the plant architecture.

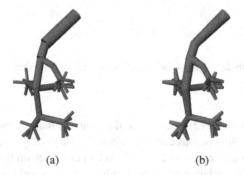

(a)   (b)

Fig. 9. This figure shows in a simplified example the effects of two different modeling approaches. Left: Chain of cylinders of decreasing radius. Right: Frustums with spheres at intersections. Ob-viously, the frustum-based approach generates a more realistic look.

The model presented here is still a prototype and has to be extended to capture the full structure of stem skeletons. Additionally the domain knowledge generated by the skeletonization (Sect. 3.3) has to be incorporated. Also, this model does not result in a photorealistic rendering of the plant but a more simplistic one, but this is no concern, since we aim at a precise structural representation of the plant architecture for deriving phenotypic features.

4.3   Volume Generation

As stated before we want to derive phenotypical features from the reconstructed geometrical structure. So far we only have a set of intersecting geometric primitives, not "real" objects. Especially we have frustums that intersect those spheres that are regarded as berries and we have frustums that intersect spheres at joints. Therefore we have to combine primitives to complex objects on the one hand and remove intersecting volumes on the other. This gives rise to two challenges: (1) We must be able to perform such operations on the primitives, (2) we must be able to detect, when to apply which operation.

One solution for challenge (1) are the well known Boolean operations *union*, *intersection* and *difference*. Since GroIMP does not yet fully support those operations we had to implement them. The Boolean operations take as input sets of triangles describing the surface of the primitives and return a set of triangles describing the surface of the complex object. Figure 10(a) shows an example of a joint consisting of two frustums and one sphere. Figure 10(b) depicts the union of the three primitives.

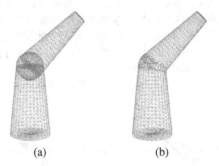

(a)   (b)

Fig. 10. Application of Boolean operations. (a) shows a joint with two frustums and one sphere and (b) its union.

For a solution to challenge (2), we can use the relations that we employ in the RGG model of the plant. Those relations can not only tell which components are connected and how, but also what Boolean operation should be performed on the connected components. For example, the relation *connected_to* for a frustum and a joining sphere can tell us that those two primitives should be unified. In contrast, the relation *attached_to* for a frustum and a sphere (a "berry sphere") tells us that we have to subtract the intersection of the two primitives from the sphere to get the correct volume of the berry.

Challenge (1) has been solved by implementing the needed Boolean operations while (2) still has to be incorporated into the RGG model and then into the Boolean operations.

5 Conclusion and Future Work

We have presented the concept of a 3D reconstruction approach of plant architecture based on the interpretation of three-dimensional laser scan point clouds. Our main idea is to use an intelligent sampling method and to constrain the space of all hypotheses by incorporating domain knowledge. As a sampler we use the Reversible Jump Markov Chain Monte Carlo method and for the modeling we rely on Relational Growth Grammars. As an example, we investigate stem skeletons of grapevine bunches. Furthermore, we will derive phenotypical features based on the reconstruction of the plants. In the domain of grapevine breeding this can, for example, be used to detect whether a newly bred cultivar shows a susceptibility to diseases like Botrytis cinerea.

For the future of this project we are planning several enhancements. For example, we want to broaden the input by using semantic surface meshes. The surface representation is based on a triangulation, generated from laser scan point clouds or point clouds created by stereo images. Then, the surface is augmented with semantic information from an image segmentation method leading to a semantic surface mesh. Additional to the knowledge modeled as an RGG this gives us further geometric and semantic constraints.

A further goal is the implementation of anaylsis tools for the structures generated by the grammar, like the total weight of a bunch or its compactness.

For the more distant future we aim at a general approach for the reconstruction of arbitrary grapevine cultivars and all of their parts, not only the bunches. Based on this, the aim is a general reconstruction approach for arbitrary plants.

Acknowledgements. This work was largely done within the sub-project D2 "Interpreted Plant Architecture" as part of CROP.SENSe.net, the German network of competence for phenotyping science in plant breeding and crop management. We thank the Federal Ministry of Science and Research (BMBF) and the European Regional Development Fund (ERDF) for funding. We thank all partners of sub-project D2 for worthful discussions. Especially, we thank Reinhard Töpfer and Katja Herzog from the Julius-Kühn-Institute, Siebeldingen, Germany for providing the stem skeletons. We thank Heiner Kuhlmann and Stefan Paulus from the Department of Geodesy of the University of Bonn, Germany for generating the laser rangefinder measurements.

References

1. Binney, J., Sukhatme, G.S.: 3D tree reconstruction from laser range data. In: International Conference on Robotics and Automation, pp. 1321–1326. IEEE (2009)
2. Bucksch, A.K.: Revealing the Skeleton from Imperfect Point Clouds. Ph.D. thesis, Delft University of Technology (2011)
3. Cornea, N., Silver, D., Min, P.: Curve–skeleton properties, applications, and algorithms. IEEE Transactions on Visualization and Computer Graphics 13(3), 530–548 (2007)

4. CROP.SENSe.net, http://www.cropsense.uni-bonn.de/
5. Ganster, B., Klein, R.: An integrated framework for procedural modeling. In: Spring Conference on Computer Graphics, pp. 150–157. Comenius University, Bratislava (2007)
6. Green, P.J.: Reversible jump Markov chain Monte Carlo computation and Bayesian model determination. Biometrika 82(4), 711–732 (1995)
7. Keightley, K.E., Bawden, G.W.: 3D volumetric modeling of grapevine biomass using Tripod LiDAR. Computers and Electronics in Agriculture 74(2), 305–312 (2010)
8. Kniemeyer, O.: Design and Implementation of a Graph Grammar Based Language for Functional-Structural Plant Modelling. Ph.D. thesis, Technical University Cottbus (2008)
9. Kurth, W., Kniemeyer, O., Buck-Sorlin, G.: Relational Growth Grammars – A Graph Rewriting Approach to Dynamical Systems with a Dynamical Structure. In: Banâtre, J.-P., Fradet, P., Giavitto, J.-L., Michel, O. (eds.) UPP 2004. LNCS, vol. 3566, pp. 56–72. Springer, Heidelberg (2005)
10. Lindenmayer, A.: Mathematical models for cellular interactions in development I. Filaments with one-sided inputs. Journal of Theoretical Biology 18(3), 280–299 (1968)
11. Lindenmayer, A.: Mathematical models for cellular interactions in development II. Simple and branching filaments with two-sided inputs. Journal of Theoretical Biology 18(3), 300–315 (1968)
12. Louarn, G., Dauzat, J., Lecoeur, J., Lebon, E.: Influence of trellis system and shoot positioning on light interception and distribution in two grapevine cultivars with different architectures: An original approach based on 3D canopy modelling. Australian Journal of Grape and Wine Research 14(3), 143–152 (2008)
13. Louarn, G., Lecoeur, J., Lebon, E.: A three-dimensional statistical reconstruction model of grapevine (Vitis vinifera) simulating canopy structure variability within and between cultivar/training system pairs. Annals of Botany 101(8), 1167–1184 (2008)
14. Müller, P., Wonka, P., Haegler, S., Ulmer, A., Van Gool, L.: Procedural modeling of buildings. ACM Transactions on Graphics 25(3), 614–623 (2006)
15. Pallas, B., Loi, C., Christophe, A., Cournède, P.H., Lecoeur, J.: A stochastic growth model of grapevine with full interaction between environment, trophic competition and plant development. In: International Symposium on Plant Growth Modeling, Simulation, Visualization and Applications, pp. 95–102. IEEE (2009)
16. Prusinkiewicz, P., Lindenmayer, A.: The Algorithmic Beauty of Plants. Springer (1990)
17. Schnabel, R., Degener, P., Klein, R.: Completion and reconstruction with primitive shapes. Eurographics 28(2), 503–512 (2009)
18. Shavrukov, Y.N., Dry, I.B., Thomas, M.R.: Inflorescence and bunch architecture development in Vitis vinifera L. Australian Journal of Grape and Wine Research 10(2), 116–124 (2004)
19. Shlyakhter, I., Rozenoer, M., Dorsey, J., Teller, S.: Reconstructing 3D tree models from instrumented photographs. IEEE Computer Graphics and Applications 21, 53–61 (2001)
20. Stiny, G., Gips, J.: Shape grammars and the generative specification of painting and sculpture. In: The Best Computer Papers of 1971, pp. 125–135. Auerbach Publications (1972)
21. Weihl, T.: Ein Identifikationsverfahren für Rebsorten. Ph.D. thesis, Universität Hohenheim (1999)

Generating Graph Transformation Rules from AML/GT State Machine Diagrams for Building Animated Model Editors

Torsten Strobl and Mark Minas

Universität der Bundeswehr München, Germany
{Torsten.Strobl,Mark.Minas}@unibw.de

Abstract. Editing environments which feature animated illustrations of model changes facilitate and simplify the comprehension of dynamic systems. Graphs are well suited for representing static models and systems, and graph transformations are the obvious choice for implementing model changes and dynamic aspects. In previous work, we have devised the Animation Modeling Language (AML) as a modeling approach on a higher level. However, AML-based specification could not yet be translated into an implementation automatically. This paper presents a language extension called AML/GT and outlines how AML/GT models can be translated into graph transformation rules automatically and also provides some implementation details.

1 Introduction

Visual models are considered to be important tools of software development. Particularly, the area of domain-specific languages (DSLs) is an interesting topic for research and industry. Therefore, some tools like *GenGEd* [3], *AToM*^{3} [6] or *DiaGen/DiaMeta* [8] support the creation of editors for DSLs with little effort. For this purpose, many of these tools generate editors out of mostly text-based editor specifications and use graphs for representing models internally, together with graph transformations (GTs) for changing them.

However, visual models are not restricted to static structures. They can also contain execution semantics, for instance the popular example of Petri nets. Editors for such models usually support animated simulations.

This paper continues our work extending the *DiaMeta* toolkit in order to facilitate the implementation of editors for complex animated models and languages with preferably minimal effort. Our first step was to allow the specification of event-based model changes through graph transformation rules (GTRs) [15]. It is based on the idea of a static graph representing an animated model that changes while time passes by. Graph transformations happen instantaneously, and they can be used for starting, stopping or modifying animations, whereas other approaches (e.g., *GenGEd* [3]) represent animations by graph transformations that do not happen instantaneously, but last as long as the animation takes.

Case studies showed that such GTR-based specifications are yet too unstructured for a convenient specification of animated systems. Therefore, in a second

A. Schürr, D. Varró, and G. Varró (Eds.): AGTIVE 2011, LNCS 7233, pp. 65–80, 2012.

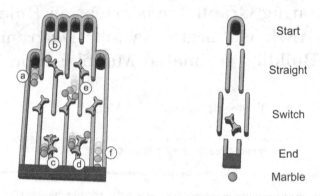

Fig. 1. Avalanche board   **Fig. 2.** Avalanche pieces

step, the modeling language AML (Animation Modeling Language) has been introduced for supporting structured design and specification of animated systems [16]. The language offers, among other features, the behavioral specification of individual components and a convenient way for describing animations for particular states. In that way, AML is a helpful tool for creating editor specifications manually, but it was not yet possible to automatically generate GTR-based specifications from AML-based specifications. This paper closes this gap and introduces AML/GT as an extension of AML as well as a tool that automatically transforms an AML/GT-based specification into a GTR-based specification for *DiaMeta* which can then be used to generate the implementation of the animated system. As a running example, an editor called Avalanche is created.

The rest of the paper is structured as follows: Section 2 outlines the running example Avalanche. Section 3 describes how animated editors can be realized with *DiaMeta* in short. AML is revisited in Section 4 and its extension AML/GT is introduced in Section 5. The translation of AML/GT state machines into GTRs is elaborated in Section 6. Section 7 gives some implementation details. Finally, Section 8 shows related work, and Section 9 concludes the paper.

2   Avalanche

This section sketches Avalanche (see Fig. 1) which has been implemented as a model editor and serves as example throughout this paper. Although Avalanche is based on a board game, gaming aspects are ignored and it is treated as an interactive dynamic system. A detailed view on Avalanche and its implementation using *DiaMeta* is presented in [15]. In addition, an animated example can be found online.[1]

In Avalanche marbles are falling down the lanes of an inclined board. The board itself can be built by four types of block pieces (see Fig. 2). The top

---

[1] http://www.unibw.de/inf2/DiaGen/animated

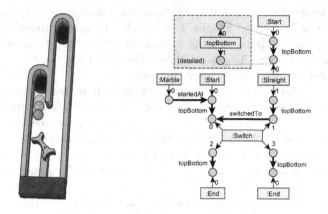

Fig. 3. AVALANCHE board and corresponding hypergraph

of each lane is limited by a *Start* piece. A *Marble* can be placed there, and it starts falling as shown in Fig. 1 (a) then. While falling, the *Marble* can be stopped by the upper side of a *Switch* (b). On the other hand, if a *Marble* hits the bottom side (c), a *Switch* is tilted to the neighboring lane. In this case, a previously stopped *Marble* can be released (d). It is also possible that a *Marble* hits another *Marble*, which is currently blocked by a *Switch*. Then, the falling *Marble* bounces off and changes its lane (e). Finally, a *Marble* can reach an *End* piece where it is removed from the board (f).

3 Specification of Animated Model Editors

AVALANCHE shall be implemented as animated model editor which allows the creation of individual boards (e.g., the board in Fig. 3). The user can put marbles in *Start* pieces. Afterwards, they start falling, and the animated model shows the game mechanics. Important aspects are the interaction between marbles and switches and that new marbles can arrive at any time, for example. The code for the AVALANCHE editor shall be generated from a specification for the *DiaMeta* system. Therefore, this section introduces some basics of *DiaMeta* and the applied generation process.

Internally, models of *DiaMeta* based editors are represented by **typed, attributed hypergraphs**. *DiaMeta* supports three types of (hyper)edges in such hypergraphs. **Component hyperedges** represent model components. These edges visit nodes that represent the component's **attachment areas**. Such areas may be spatially related, e.g., overlap, which is represented by so-called **relation edges** between the corresponding nodes. Relation edges are created or removed automatically whenever attachment areas of components start or end being spatially related [8]. Finally, **link hyperedges** can be used for connecting nodes of component hyperedges. Such edges can be created and removed explicitly.

In Fig. 3 an AVALANCHE board is shown together with its internal hypergraph. Because of such underlying graphs, it is possible to specify model changes via GTs. Our approach of implementing animated editors is also based on GTs, so a brief description follows. A detailed elaboration, however, can be found in [15], which also includes the underlying formalism.

The internal graph does not necessarily **represent a static model**. Instead, a constant graph may describe how visible components are **animated**, e.g., by determining where, when and how a component started moving. Then, the visualization of the graph can involve the current animation time for illustrating an animated scene. The hypergraph in Fig. 3 actually shows a marble with appended *startedAt* edge, which stands for a currently falling marble during the visualization of the model.

Besides such (graphical) animations, there are also **time-dependent model changes** caused by GTs. These GTs are the result of internal and external events. **External events** are sent to the system from an external source, e.g., from a user sending a command to the editor, whereas the occurrence of **internal events** is based on the system state. Such events are specified (and implemented) by GTRs, or rather graph transformation programs in *DiaMeta*. They are equipped with application conditions to trigger them when external events happen, or with a **time calculation rule (TCR)** that computes the time when the internal event occurs. This calculation usually depends on the attribute values of the related components.

However, the behavior of the AVALANCHE editor is not specified directly via GTRs and event details as described above in this paper. Instead, the modeling language AML/GT (see Sections 4 and 5) is used and GTRs will be generated.

4   Animation Modeling Language (AML)

The presented specification approach comes along with the Animation Modeling Language (AML). This section provides an overview focused on aspects relevant for this paper. Some more details, origin, and goals of AML are presented in [16].

AML models describe both the structure and relations of graphical components and their dynamic behavior including animations. The language is based on UML 2.3 [11]. AML refines UML class diagrams for the static structure and UML state machine diagrams for the dynamic part (see Fig. 4).

The following extensions are available for the static structure:

- **Media Components** are the main structural element of AML. They extend regular UML classes by aspects of visual components. They always include basic attributes like *xPos*, *yPos*, *angle* and other attributes which determine spatial aspects or drawing state.
- **Inner Properties** are similar to UML properties, but their specified type must be a media component. In addition, they can be arranged hierarchically. In this way, inner properties can model the compositions of graphical sub elements within their containing media components.

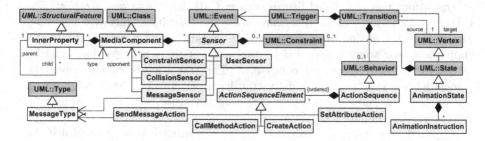

Fig. 4. AML metamodel (excerpt)

- **Sensors** are owned by media components and observe states, animations and interactions of its owner.

The three most important kinds of sensors are:

- **User Sensors** are *triggered* if the user interacts with the sensor owner in a specific way, e.g., if the user clicks on a component.
- **Constraint Sensors** are *triggered* if its guard expression (by default an OCL expression [10]) evaluates to *true*, which may also happen during an animated situation. **Collision Sensors** are a special subtype of constraint sensors. They can observe when the owner collides with another component, called **opponent**.
- **Message Sensors** are *enabled* if the sensor's owner receives a message of a declared type from another component, called **opponent**, too.

Fig. 5 shows an AML/GT model, i.e., an AML model with GT extensions described in the next section, but it illustrates plain AML elements, too. The inner property *lever* ❷,[2] e.g., represents the graphical lever within the media component *Switch* ❶ which represents the whole board piece. It is also shown how the default value of the *lever*'s attribute *angle* is redefined there. The user sensor *PutMarble* ❸ is *triggered* if the user clicks on a *Start* component, which is the sensor owner ❹ indicated by the solid line. The collision sensor *MarbleEnd* ❺ is *triggered* if its owner *Marble* collides with the bottom of an *End* piece, which is connected via a dashed **opponent arrow** ❻.

Behavior and animations of media components are described by state machine diagrams and their regions ❽. Each state machine models the states and transitions of components of one media component type.

For the description of graphical animations, AML supports special states called **animation states** ❾ that contain so-called **animation instructions** ❿. Each animation instruction computes the value of an attribute of the state machine's media component while time passes by. E.g., animation instruction ❿ defines that the angle of a *Switch* lever changes from −30° to 30° within three seconds as soon as the *Switch* enters state *LeftRight*.

---

[2] Numbers in black circles refer to the ones in Fig. 5.

Transitions can be labeled like in regular UML by *trigger[guard]/effect*. The **trigger** describes an event that may cause the transition. Intuitive keywords like *at* or *after* ⑪ indicate certain points in time or time delays. AML also supports sensor triggers specified by the name of the sensor ⑫ that must be owned by the state machine's component. The transition is *enabled* when the associated sensor is *triggered*. Finally, transitions without trigger specification ⑬ are *triggered* as soon as the state's internal activities (e.g., active animation instructions) terminate (*completion event*).

A **guard** may restrict transition triggers; a transition is only *enabled* if the guard condition is satisfied, too, like in UML.

In AML, effects as well as *entry* and *exit* behaviors of states are restricted to so-called **action sequences** consisting of an arbitrary series of the following actions:

- **Set actions** ⑭ can change an attribute of the corresponding component or one of its inner properties.
- **Create actions** ⑮ allow the creation of media components.
- **Call actions** ⑯ are used for calling operations.
- **Send actions** can send messages to other media components.

The **execution of transitions** is performed sequentially. Since a system with discrete time model may have multiple *enabled* transitions at the same time and the UML specification does not define full semantics of conflicting transitions, *enabled* AML transitions are executed with the following priority, starting with the highest one: sensor events, time events, without trigger/with guard, without trigger/without guard. In addition, each sensor can have an explicitly specified numeric priority value. Such values must be used to order a set of simultaneously occurring sensor events. If there are still conflicts, a transition must be chosen randomly.

5 Animation Modeling Language for GTs (AML/GT)

AML enables developers to model AVALANCHE and to translate it to a GTR-based *DiaMeta* specification manually as described in [16]. For an automated translation, however, AML models lack important information. Therefore, AML has been extended to AML/GT which addresses this issue. Technically, AML/GT is realized as a (UML) profile for extending AML models. This section describes some of its elements in the context of the AVALANCHE model shown in Fig. 5.

Some media components represent visual components that the user can place on the screen using the generated AVALANCHE editor. They are marked by a small square next to the media component icon, e.g., *Switch* ❶, and are then represented by *component hyperedges* (see Section 3). In addition, inner properties of media components can be marked by small circles ⑰. Such properties are represented by nodes of this component hyperedge. Inner properties which model graphical aspects of the component are marked by a small eye ❷.

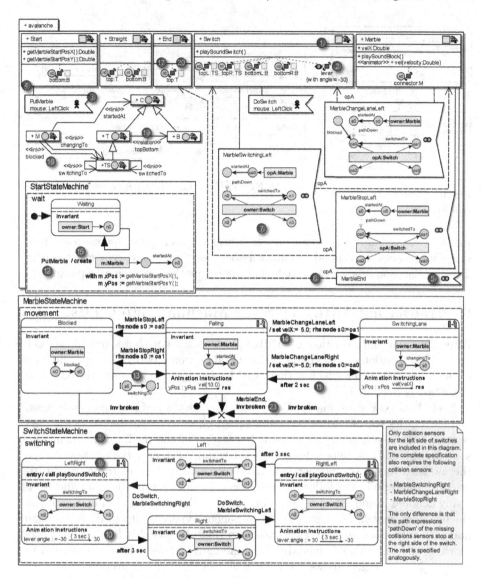

Fig. 5. Avalanche specification (AML/GT model)

A *Switch* hyperedge, for example, visits the nodes *topL*, *topR*, *bottomL* and *bottomR* which represent the four corners of the corresponding component. The types of these nodes are *TS* ("top/switch") or *B* ("bottom"), resp., as specified in Fig 5 ❶. Fig. 5 also shows how attributes of the inner property's media component type are redefined by new default values ❷.

Moreover, associations between media components can be interpreted as edge types which can be used to connect the nodes of component hyperedges. Associations corresponding to *link hyperedges* are marked with stereotype **link** ⑱.

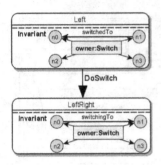

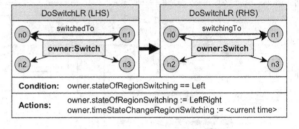

Fig. 6. Exemplary transition for *Switch*

Fig. 7. GTR for *DoSwitch* transition

Relation edges indicating spatial relationships between attachment areas of visual components correspond to binary associations with stereotype **relation** ⑲. The specification of such an attachment area is marked by a small square ⑳.

6   Translating AML/GT State Machines into GTRs

After the basic introduction of AML/GT, this section shows how AML/GT models are translated into GTRs. These GTRs become part of the specification of the animated editor which is then used by *DiaMeta* to generate the editor as described in Section 7.

Each AML/GT state machine describes the behavior of a media component, and each of its states corresponds to a subgraph of the animated diagram's hypergraph. It contains the component hyperedge of the media component and some additional link hyperedges. This subgraph represents the situation when the modeled media component is in this state, i.e., such a subgraph, called **invariant pattern** in the following, represents invariant of the state and is visually denoted inside the state box. For example, the state *Left* in Fig. 6 contains the link edge *switchedTo* next to the component hyperedge itself. By connecting node *n1* with node *n0*, the edge indicates that the switch's *lever* is currently blocking the *Switch*'s left lane. In the same figure, there is also a transition to state *LeftRight*. Fig. 7 shows the GTR realizing this state transition.

In order to create the required set of GTRs, the algorithm has to iterate over all transitions with or without triggers (a completion event trigger is assumed in the latter case). During this process, the triggers have **priorities** according to the rules at the end of Section 4. Each GTR of the resulting set can be specified with this priority then. GTs with higher priority can be processed first, if there is more than one GT scheduled at the same point in time.

The following steps show how single GTRs are created. They are based on the example in Fig. 8 and its example trigger *MarbleChangeLaneLeft* ❶. The illustrated transition is responsible for changing a *Marble*'s state from *Falling* to *SwitchingLane*, i.e., the *Marble* has to move to another lane because it has hit the left side of a *Switch* which is currently blocked by another *Marble*. The graph

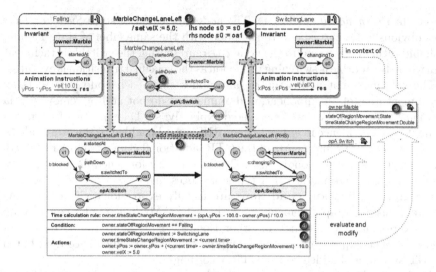

Fig. 8. Generating a GTR from a state machine transition

pattern shown within the collision sensor *MarbleChangeLaneLeft* represents this condition.

The **LHS** of the rule is created first. It consists of the invariant pattern of the transition's source state, extended by subgraphs of the transition's guard (e.g., Fig. 5 ⑬) as well as graph patterns describing the behavior of sensors being used as transition triggers (e.g., Fig. 5 ❼). In the example shown in Fig. 8, there is no transition guard, but the graph pattern located within collision sensor *MarbleChangeLaneLeft* has to be added. In general, the LHS is constructed from several graph patterns. Negative application conditions are possible, too, but they are not required here.

As a basic principle, the creation of the LHS from all these graph patterns is performed by building the union of them and then gluing them in a suitable way. Gluing is rather straight forward because each of the graph patterns refers to the same "owner" of the state machine which is represented by its component hyperedge and its visited nodes. Therefore, the LHS may contain the component hyperedge of its owner only once, i.e., all instances of this hyperedge and its visited nodes must be glued. Further nodes can be glued as well. In the example, node *s0* of the source state graph pattern must be glued with *s0* of the pattern in the sensor. This is specified by the **correspondence statement** *lhs node s0 := s0* ❷. However, because equally labeled nodes are glued automatically, such a statement is omitted in Fig. 5.

The **RHS** of the rule is built in a similar way. It is constructed from the invariant of the transition's target state and, again, all graph patterns of the transition as for the LHS. The latter have to be added, too, because applying the rule shall just change the state of the owner's state machine, i.e., change its invariant pattern; the rest of the diagram's hypergraph must remain unchanged.

The owner's component hyperedge again determines which edges and nodes must be glued. The gluing of an additional node is specified by the correspondence statement *rhs node s0 := oa1* ❷. The node labeled *s0* of state *SwitchingLane* does obviously not correspond to the node with the same label of state *Falling*; it should rather correspond to the node *oa1* representing the switch's attachment area where the marble is switching to as soon as the specified transition is executed.

The rule must not add or delete any nodes since each node represents an attachment area of a component, and animations do not add or remove media components (except if created or deleted explicitly; see below). Therefore, nodes without correspondence in either LHS or RHS must be added to the other side ❸.

After creating the main parts of the GTR, additional elements must be added, i.e., further application conditions, a TCR necessary for scheduling events, attribute changes, and maybe other types of processing.

Further application conditions are usually expressions which check attribute values of graph elements. For this, the AML/GT model can contain OCL expressions (conditions), or other Boolean expressions (Java) in case of *DiaMeta*. Such expressions must be adapted to a syntax which is compatible to the GT system and added to the GTR accordingly.[3]

As described above, the source state's invariant pattern is part of the corresponding GTR's LHS. However, just relying on the invariant pattern of the source state is generally not sufficient as the *inv broken* trigger shows (see below). Therefore, a **state attribute** is added to the component hyperedge ❺ which is checked before the rule may be applied ❹. There must be one such attribute for each UML region because of concurrent and hierarchical states.

Another type of application conditions are **path expressions**. In Fig. 8, a path expression called *pathDown* is required ❻. It verifies that the *Marble* is currently falling down the lane which leads through the blocked left side of the *Switch*. In *DiaMeta* this expression is specified as an arbitrary sequence of *topBottom(0,1)*, *Switch(0,2)*, *Switch(1,3)*, or *Straight(0,1)* edges. The numbers within the parenthesis specify the hyperedge tentacles the path must follow: the first number specifies the ingoing tentacle and the second one specifies the outgoing tentacle when following the path through the hypergraph.

AML/GT allows specifying **actions** at different places. They may be specified with each transition; they are executed as soon as the transition fires (e.g., Fig. 5 ⓮) or as an action associated with an entry into a state (e.g., Fig. 5 ❾). Actions are easily translated into GTRs in *DiaMeta* since it allows arbitrary Java code when executing a rule. *Call actions* result in the calls of media component operations (generated Java code; e.g., Fig. 5 ❾), and *set actions* involve the change of attributes of the component hyperedge during the GT (e.g., Fig. 5 ⓮). Finally, *create actions* can construct components (e.g., Fig. 5 ⓯): AML/GT allows declaring component hyperedges (here *Marble*) including its initial attributes, its nodes, and link edges. These edges and nodes must be added to the RHS of the GTR, and attributes must be set accordingly.

---

[3] Single graph transformation rules are not sufficient in the following; graph transformation programs are actually needed. However, for simplicity, we still use the term "graph transformation rule" (GTR) instead of "graph transformation program".

Further actions must be added to each GTR (Fig. 8 ❼). First, the *state attribute* and the **state entry time** must be updated (such an attribute must be available for each region again). And second, the attributes which are considered *animated* during an active animation state must be updated because attributes do not change their values during animations in our approach (cf. Section 3). Instead, the changing value is calculated using a formula considering the animation time. Therefore, an update which applies the last calculated value is required.

A generated GTRs must be executed at a specific point in time which is modeled by a transition's **trigger**. A transition triggered by a user sensor is translated into a GTR that can be induced by the user in the editor. By default, the GTR is bound to a GUI button starting the GTR if it is applicable. Other sensor triggers require a **TCR** ❽ for scheduling the GT (see Section 3).

Transitions with time event triggers are translated into GTRs representing *internal events* with a TCR reflecting the absolute or relative time when the event may be triggered. Relative time always refers to the *state entry time*.

Transitions without explicit trigger must be executed as soon as the pattern of the LHS can be matched and animations of the source state have been finished (an additional condition). They are translated into GTRs representing *internal events* as well. Its TCR must return the point in time when animations have been finished. If there is no animation, the GTR must be applied without any time delay, i.e., the TCR always returns the current time.

Constraint and collision sensors result in internal event specifications, too. They usually check *animated* attributes (see above). However, generating a corresponding TCR is not straight forward. For a collision sensor, e.g., a TCR is required which calculates the collision time based on the components' trajectories. Currently, such a TCR has to be provided manually. Using a physics engine may solve this problem in the future.

Finally, two special cases for generating GTRs from the Avalanche model are described. **Termination states** are used as the only means to delete component hyperedges and, therefore, media components. Such states are visualized as a small X. If a transition ends in a termination state, it is translated into a GTR that removes the corresponding component hyperedge and its nodes, so the resulting RHS corresponds to the LHS without the component's edge and its nodes.

Inv broken is an AML/GT keyword that may be used as a transition trigger. It has lowest priority and is *fired* if the transition's source state must be left because its invariant pattern is "broken". This may happen if some link edges being part of the invariant pattern are deleted when some other component and its link edges are deleted due to a user action.

For instance, a *Switch* may be removed by the user even if a *Marble* is connected to one of its nodes via *startedAt* (state *Falling* in Fig. 5 ㉑). Because the specification in Fig. 5 requires a marble to be connected to the component where it started falling, the *Marble* would be in an inconsistent state then. The *inv broken* trigger is used to represent the fact that the pattern invariant of the current state is suddenly no longer true although the state has not yet been left.

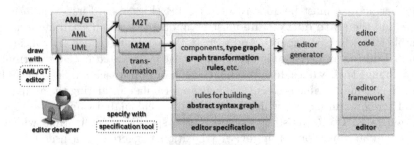

Fig. 9. Overview of the editor generation process

This triggers the state's *inv broken* transition and activates its termination state which removes the *Marble*, too.

The translation of such an *inv broken* transition into a GTR is straight forward: The GTR must check the component hyperedge's *state attribute* (see above) and whether the invariant pattern is violated. The latter is simply represented as a negative context, i.e., the GTR may be applied if the component is in the corresponding state, but its invariant pattern can (no longer) be matched.

In order to support all features of AML/GT state machines, further special cases must be considered when generating GTRs as well. Namely, **composite states**, **concurrency** and **message passing** must be processed specifically. These topics are not discussed here, but it is clear that they can be translated to GTRs, too.

7 Implementation

The translation process outlined in the previous section has been completely implemented, also covering those topics that have been omitted here, e.g., composite states, orthogonal regions or message passing. It is possible now to specify animated editors (e.g., for AVALANCHE) using a visual specification tool (the AML/GT editor) and to generate the editor from this specification. This section sketches the process, which is outlined in Fig. 9, and some design decisions.

For the creation of AML/GT models, the editor designer uses the AML/GT editor (see Fig. 10), which has been generated using the *DiaMeta* toolkit, too. AML/GT models are automatically translated into the specified animated editor. Some Java code of the editor is directly generated from the AML/GT model ("M2T" in Fig. 9). This is necessary for support code which has to be provided manually when using *DiaMeta* without AML/GT. The "M2T" transformation has been realized with *Acceleo*, a Model-to-Text translation language which is oriented towards the **MOFM2T** specification of the OMG [9].

The translation process from an AML/GT model into an editor specification for *DiaMeta* ("editor specification" in Fig. 9) as outlined in the previous section is performed by the "M2M" component in Fig. 9. It uses *Eclipse QVTo*, which implements the OMG specification of **MOF QVT Operational** [12],

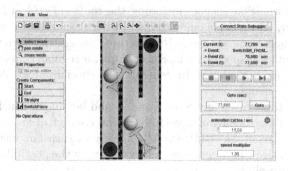

Fig. 10. Screenshot of the AVALANCHE editor

a Model-to-Model technology. The target model of this transformation is the native, XML-oriented *DiaMeta* specification which also includes the GTRs.

There are several reasons for choosing these technologies instead of using the *DiaMeta* framework for creating the specification from AML/GT models. First of all, the GT system of *DiaMeta* has been created for structured editing in editors and therefore lacks convenient features for complex model transformations. In addition, both source (AML/GT) and target model (*DiaMeta* specification) do not consist of graphs only, but also contain embedded text or must produce Java code which are less suited for being modeled as graphs. Finally, standardized languages have been favored in order get a more future-proof, general and stable transformation. The mentioned languages have already been supported by a couple of tools (Eclipse plugins), too.

Besides the AML/GT model, the editor designer has to provide the abstract syntax[4] of the animated visual language, i.e., its meta-model. *DiaMeta* simply uses EMF for metamodelling. Any EMF-based tool for specifying Ecore models [14] can be used here. Furthermore, generating TCRs from collision sensors still needs manual code (see Section 6), so future solutions require a generic *a priori* collision detection algorithm.

Finally, the "editor generator" creates the Java code of the animated editor using the code generator of *DiaMeta*. The screenshot in Fig. 10 shows the generated AVALANCHE editor with created AVALANCHE board and a falling marble. This editor has been generated from the AML/GT model shown in Fig. 5.

8   Related Work

Although it has not been the primary intention of AML/GT, it can be considered as a visual language for programmed graph rewriting and model transformation. A good starting point for reading into this topic is provided in [4], which compares *AGG*, *Fujaba*, and *PROGRES* as graph transformation languages and also mentions *GReAT*. Another overview is presented in [18] which focuses

---

[4] All descriptions in this paper and AML/GT concern the concrete syntax only.

on model transformation aspects including the tools $AToM^3$, $VIATRA2$, and $VMTS$. However, AML/GT cannot compete with these rather general-purpose languages and tools because the GTRs generated are restricted to the very specific field of animation specification. For example, a regular transition can only add or remove link hyperedges. In general, it is specialized for describing the behavior of components.

Furthermore, there are many other systems and languages which are also intended for describing behavior, especially in the context of statecharts, UML, and MDE, e.g., Executable UML (xUML). Many of such languages try to extend or constrain UML in order to get a language with the possibility to describe systems more precisely than UML. However, many of them have a specific application domain. AML and AML/GT, which also extend and constrain standard UML, focus on visual appearance and graphical animations next to the expressiveness of UML statecharts and hypergraphs. Therefore, they are especially suited for specifying animated visual languages and generating editors. To the best of our knowledge, they are the only languages featuring this combination, so the following related work has more or less other domains.

Fujaba, for example, supports so-called story diagrams [2] which combine UML collaboration diagrams with activity diagrams. GTRs can be drawn within activities which allow the creation of complex transformation flows. Originally, the main purpose was to generate Java code from such models, but areas of application evolved and many extensions are available in the meantime. For example, *Fujaba Real-Time* [1] allows the modeling of embedded systems based on statecharts with real-time capabilities. The modeling of visual aspects of components, however, is not provided. On the other side, AML/GT is not suited for embedded systems and does not provide the modeling of real-time constraints.

Another modular and hierarchical model-based approach is presented in [17]. The semantic domain of the models presented there is the Discrete Event System Specification ($DEVS$). It is used for describing control structures for programmed graph rewriting. Although the formalism has a solid foundation, it requires the user to have a specialized expertise while UML-based approaches are well-known to the majority of all users. Moreover, it also lacks the possibility to define the structure of components or graphic related issues.

In [13] a visual language for model transformations and specifying model behavior has been introduced which allows the specification of in-place transformation rules. These rules can be compared to GTRs in a graph-based environment. There are some further extensions like rule periodicity, duration, exceptions, etc. State-based modeling on a more coherent level, which is also the intention of this paper, is not provided.

Finally, the animation approach used in this paper and the translation of state machines into GTRs is related to other approaches. The differences between our animation approach and those approaches in similar systems (e.g., [3]) have been discussed in previous work [15] already. Translating state machines into equivalent GTR systems is not a new idea either. E.g., there are several papers describing the semantics of (UML) state machines based on GTs, e.g., in [5].

The translation process described in Section 6 is rather tailored to its specific application within animated editors. Finally, the way graph patterns are used in state machines and sensors of AML/GT can be considered as alternative to OCL [10]. We also would like to point towards agent-based modeling [7] as another field of application for AML/GT.

9 Conclusions

We have pursued our approach of modeling animated editors with the UML-based language AML. The language extension AML/GT offers additional elements which are necessary to create specifications of dynamic systems in a hypergraph-based environment automatically.

Applying a modeling language like AML/GT promises that complex systems can be specified in a clear and accessible way. Using state machines for individual components particularly complies with an intuitive perspective on many systems. At the same time, the underlying GTs provide an established foundation and exact semantics, and they facilitate the comprehension of the execution process.

The algorithmic translation of AML/GT model into the specification format of *DiaMeta* has been completely implemented. The aim of creating a higher-level modeling language, which does not require further specifications for generating interactive, animated editors, has been accomplished. Future work will concentrate on how collisions between animated components can be detected without computing their trajectories in advance. Using a physics engine as used in many game settings appears to be promising.

References

1. Burmester, S., Giese, H.: The Fujaba Real-Time Statechart Plugin. In: Giese, H., Zündorf, A. (eds.) Proc. of the 1st International Fujaba Days 2003, pp. 1–8 (2003); Technical Report tr-ri-04-247, Universität Paderborn, Informatik
2. Diethelm, I., Geiger, L., Zündorf, A.: Systematic Story Driven Modeling, a Case Study. In: Giese, H., Krüger, I. (eds.) Proc. of the 3rd International Workshop on Scenarios and State Machines: Models, Algorithms, and Tools (SCESM 2004), ICSE Workshop (2004)
3. Ermel, C.: Simulation and Animation of Visual Languages Based on Typed Algebraic Graph Transformation. Ph.D. thesis, Technical University Berlin (2006)
4. Fuss, C., Mosler, C., Ranger, U., Schultchen, E.: The Jury is Still Out: A Comparison of AGG, Fujaba, and PROGRES. In: Ehrig, K., Giese, H. (eds.) Proc. of the 6th International Workshop on Graph Transformation and Visual Modelling Techniques. ECEASST, vol. 6 (2007)
5. Kuske, S.: A Formal Semantics of UML State Machines Based on Structured Graph Transformation. In: Gogolla, M., Kobryn, C. (eds.) UML 2001. LNCS, vol. 2185, pp. 241–256. Springer, Heidelberg (2001)
6. de Lara, J., Vangheluwe, H.: AToM³: A Tool for Multi-formalism and Meta-modelling. In: Kutsche, R.-D., Weber, H. (eds.) FASE 2002. LNCS, vol. 2306, pp. 174–188. Springer, Heidelberg (2002)

7. Macal, C.M., North, M.J.: Tutorial on agent-based modeling and simulation. In: Kuhl, M.E., Steiger, N.M., Armstrong, F.B., Joines, J.A. (eds.) Proc. of the 37th Winter Simulation Conference, pp. 2–15. ACM (2005)

8. Minas, M.: Generating Meta-Model-Based Freehand Editors. In: Zündorf, A., Varró, D. (eds.) Proc. of the 3rd International Workshop on Graph-Based Tools. ECEASST, vol. 1 (2006)

9. Object Management Group (OMG): MOF Model To Text Transformation Language, v1.0 (January 2008), http://www.omg.org/spec/MOFM2T/1.0

10. Object Management Group (OMG): Object Constraint Language, v2.2 (February 2010), http://www.omg.org/spec/OCL/2.2

11. Object Management Group (OMG): Unified Modeling Language: Superstructure, v2.3 (May 2010), http://www.omg.org/spec/UML/2.3/Superstructure

12. Object Management Group (OMG): MOF Query/View/Transformation, v1.1 (January 2011), http://www.omg.org/spec/QVT/1.1

13. Rivera, J.E., Durán, F., Vallecillo, A.: A Graphical Approach for Modeling Time-Dependent Behavior of DSLs. In: DeLine, R., Minas, M., Erwig, M. (eds.) Proc. of the 2009 IEEE Symposium on Visual Languages and Human-Centric Computing, pp. 51–55. IEEE Computer Society (2009)

14. Steinberg, D., Budinsky, F., Paternostro, M., Merks, E.: EMF: Eclipse Modeling Framework, 2nd edn. Addison-Wesley (2009)

15. Strobl, T., Minas, M.: Specifying and Generating Editing Environments for Interactive Animated Visual Models. In: Küster, J., Tuosto, E. (eds.) Proc. of the 9th International Workshop on Graph Transformation and Visual Modeling Techniques. ECEASST, vol. 29 (2010)

16. Strobl, T., Minas, M., Pleuß, A., Vitzthum, A.: From the Behavior Model of an Animated Visual Language to its Editing Environment Based on Graph Transformation. In: de Lara, J., Varró, D. (eds.) Proc. of the 4th International Workshop on Graph-Based Tools. ECEASST, vol. 32 (2010)

17. Syriani, E., Vangheluwe, H.: Programmed Graph Rewriting with Time for Simulation-Based Design. In: Vallecillo, A., Gray, J., Pierantonio, A. (eds.) ICMT 2008. LNCS, vol. 5063, pp. 91–106. Springer, Heidelberg (2008)

18. Taentzer, G., Ehrig, K., Guerra, E., de Lara, J., Lengyel, L., Levendovszky, T., Prange, U., Varró, D., Varró-Gyapay, Sz.: Model Transformation by Graph Transformation: A Comparative Study. In: Proc. Workshop Model Transformation in Practice (Satellite Event of MoDELS 2005) (2005)

AGG 2.0 – New Features for Specifying and Analyzing Algebraic Graph Transformations

Olga Runge[1], Claudia Ermel[1], and Gabriele Taentzer[2]

[1] Technische Universität Berlin, Germany
{olga.runge,claudia.ermel}@tu-berlin.de
[2] Philipps-Universität Marburg, Germany
taentzer@informatik.uni-marburg.de

Abstract. The integrated development environment AGG supports the specification of algebraic graph transformation systems based on attributed, typed graphs with node type inheritance, graph rules with application conditions, and graph constraints. It offers several analysis techniques for graph transformation systems including graph parsing, consistency checking of graphs as well as conflict and dependency detection in transformations by critical pair analysis of graph rules, an important instrument to support the confluence check of graph transformation systems. AGG 2.0 includes various new features added over the past two years. It supports the specification of complex control structures for rule application comprising the definition of control and object flow for rule sequences and nested application conditions. Furthermore, new possibilities for constructing rules from existing ones (e.g., inverse, minimal, amalgamated, and concurrent rules) and for more flexible usability of critical pair analyses have been realized.

Keywords: graph transformation tool, AGG 2.0.

1 Introduction

AGG [15,13] is a well-established tool environment for algebraic graph transformation systems, developed and extended over the past 20 years. Graphs in AGG are defined by a type graph with node type inheritance and may be attributed by any kind of Java objects. Graph transformations can be equipped with arbitrary computations on these Java objects described by Java expressions.

The AGG environment consists of a graphical user interface comprising several visual editors, an interpreter, and a set of validation tools. The interpreter allows the stepwise transformation of graphs as well as rule applications as long as possible. AGG supports several kinds of validations which comprise graph parsing, consistency checking of graphs, applicability checking of rule sequences, and conflict and dependency detection by critical pair analysis of graph rules. Applications of AGG include graph and rule-based modeling of software, validation of system properties by assigning an operational semantics to some system model, graph transformation-based evolution of software, and the definition of visual languages by graph grammars.

A. Schürr, D. Varró, and G. Varró (Eds.): AGTIVE 2011, LNCS 7233, pp. 81–88, 2012.

Model transformations have recently been identified as a key subject in model-driven development (MDD). Graph transformations offer useful concepts for MDD, while the software engineering community can generate interesting challenges for the graph transformation community. In the past two years, those new challenges also lead to the augmentation of AGG by new tool features. In this paper, we describe some of those challenges, the formal approaches developed to solve them, and the impact they had on recent developments of new features of AGG, leading to AGG 2.0.

2 Rule Application Control

Application Conditions. For graph transformation rules, well-known negative or positive application conditions (NACs, PACs) may be used that forbid or require a certain structure to be present in the graph for the rule to be applied. As a generalization, *application conditions* (introduced as *nested application conditions* in [9]) further enhance the expressiveness of graph transformations by providing a more powerful mechanism to control rule applications (see [8] for an extensive case study).

An *application condition ac* of rule $r : L \to R$ is of the form *true* or $\exists(a, c)$ where $a : L \to C$ is a graph morphism from L to a condition graph C, and c is a condition over C. Application conditions may be nested, negated,[1] quantified by FORALL and combined by using the logical connectors AND and OR. Given application condition ac, a match $m : L \to G$ satisfies ac, written $m \models ac$, if $ac = true$. A match $m : L \to G$ satisfies condition $ac = \exists(a, c)$ if there is an injective graph morphism $q : C \to G$ such that $q \circ a = m$ and q satisfies c. The satisfaction of conditions is extended to Boolean conditions in the usual way. A rule $L \to R$ is applicable only if the application condition ac is satisfied for its match $m : L \to G$.

An example is shown in Figure 1, where the activation of an elementary Petri net transition is checked: Rule ActivationCheck sets the transition attribute isActivated to true if two conditions, called PreCond and PostCond are satisfied: PostCond is the NAC shown in Figure 1 (b) which forbids the existence of a marked place in the transition's post-domain, and PreCond is a nested application condition shown in Figure 1 (c) which requires that on each place in the transition's pre-domain, there must be one token. Note that this condition cannot be expressed by using simple NACs or PACs. Figure 1 (a) shows the context menu entries for generating general application conditions (GACs) in AGG 2.0.

Object Flow for Rule Sequences. Object flow between rules has been defined in [11] as partial rule dependencies relating nodes of the RHS of one rule to (type-compatible) nodes of the LHS of a (not necessarily direct) subsequent rule in a given rule sequence. Object flow thus enhances the expressiveness of graph transformation systems and reduces the match finding effort.

---

[1] An application condition of the form $\neg \exists a$ is called *negative application condition*.

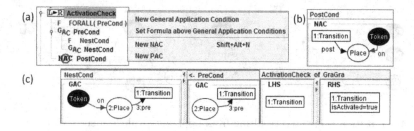

Fig. 1. Nested application conditions in AGG

In AGG 2.0, object flow can be defined between subsequent rules in a rule sequence, and the rule sequence can be applied to a given graph respecting the object flow. An example is the definition of a Petri net transition firing step by the rule sequence (ActivationCheck, RemovePre*, AddPost*, DeActivate) with object flow. The sequence defines that rule ActivationCheck (see Figure 1) is applied once, followed by rules RemovePre (removing a token from a pre-domain place), AddPost (adding a token to a post-domain place), which are shown in Figure 2, and DeActivate (setting the transition attribute isActivated back to false).

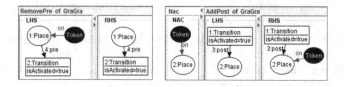

Fig. 2. Rules RemovePre and AddPost for Petri net firing

In the sequence, rules RemovePre and AddPost are applied as long as possible (denoted by "*"), and rule DeActivate is applied once. To restrict the application of the rule sequence to exactly one transition, we need to express that the transition in the matches of all rules is the same. This is done by defining the object flow, e.g., by mapping the transition from the RHS of rule ActivationCheck to the LHSs of rules RemovePre and AddPost as shown in Figure 3.

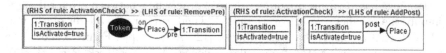

Fig. 3. Object flow definition for rule sequences in AGG

An example of a firing step by applying the rule sequence with object flow is shown in Figure 4, where the left transition in the net was selected, found activated and fired.

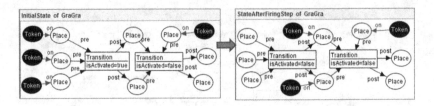

Fig. 4. Transition firing step resulting from applying a rule sequence with object flow

3   Constructing Rules from Existing Ones

Obviously, the definition of a rule sequence with object flow is neither simple nor very intuitive to model Petri net transition firing steps for transitions with an arbitrary number of pre- and post-domain places. A way closer to the inherent Petri net semantics (a transition removes the tokens from all of its pre-domain places and adds tokens to all of its post-domain places in *one atomic* step) would be to construct a *single* rule modelling the firing behaviour of a transition. AGG 2.0 supports various ways to automatic rule construction.

Construction of Concurrent Rules. A concurrent rule summarizes a given rule sequence in one equivalent rule [5,12]. In a nutshell, a concurrent rule $p *_E q = (L \leftarrow K \rightarrow R)$ is constructed from two rules $p = (L_p \leftarrow K_p \rightarrow R_p)$ and $q = (L_q \leftarrow K_q \rightarrow R_q)$ that may be sequentially dependent via an overlapping graph E by the diagram to the right, where (5) is pullback (intersection of graphs C_p and C_q in E), and all remaining squares are pushouts. (A pushout is a union of two graphs over an interface graph.) The concurrent rule $(p_1 * \ldots * p_n)_E$

$$L_p \leftarrow K_p \rightarrow R_p \qquad L_q \leftarrow K_q \rightarrow R_q$$
$$\downarrow (1) \downarrow \qquad (2) \searrow {}^{e_p} {}^{e_q} \swarrow \qquad (3) \downarrow (4) \downarrow$$
$$L \leftarrow C_p \longrightarrow E \longleftarrow C_q \rightarrow R$$
$$(5)$$
$$K$$

with $E = (E_1, \ldots, E_{n-1})$ for a sequence is constructed in an iterated way by $(p_1 * \ldots * p_n)_E = p_1 *_{E_1} p_2 *_{E_2} \ldots *_{E_{n-1}} p_n$. In AGG 2.0, we can construct a concurrent rule from a rule sequence with the following options concerning the overlappings of one rule's RHS with the succeeding rule's LHS: (1) compute maximal overlappings according to rule dependencies; (2) compute all possible overlappings (this usually yields a large number of concurrent rules); (3) compute overlappings based on the previously defined object flow between given rules; (4) compute the parallel rule (no overlapping), where rule graphs are disjointly unified, with NACs constructed according to [12]. As example, Figure 5 shows the concurrent rule constructed from rule sequence (ActivationCheck, RemovePre(3), AddPost(2), DeActivate).[2] Since an object flow is defined for this sequence, we choose option (3) *by object flow* for computing rule overlappings. Constraints on the type graph (e.g., "a Token node is connected to exactly one Place node") prevent the generation of unnecessary NACs.

---

[2] Note that a concurrent rule can be constructed only for a finite rule sequence.

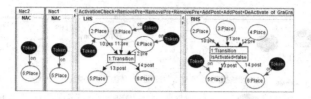

Fig. 5. Concurrent rule generated from a rule sequence in AGG

Construction of Amalgamated Rules. If a set of rules $p_1, \ldots, p_n$ share a common subrule p_0, a set of "amalgamable" transformations $G \overset{(p_i, m_i)}{\Longrightarrow} G_i$ $(1 \leq i \leq n)$ leads to an amalgamated transformation $G \overset{\tilde{p}, \tilde{m}}{\Longrightarrow} H$ via the amalgamated rule $\tilde{p} = p_1 +_{p_0} \ldots +_{p_0} p_n$ constructed as gluing of $p_1, \ldots, p_n$ along p_0. We call p_0 *kernel rule*, and $p_1, \ldots, p_n$ *multi-rules*. A kernel rule together with its embeddings in several multi-rules is called *interaction scheme*. An amalgamated transformation $G \overset{\tilde{p}}{\Longrightarrow} H$ is a transformation via the amalgamated rule $\tilde{p}$ [8].

This concept is very useful to specify ∀-quantified operations on recurring graph patterns (e.g., in model refactorings). The effect is that the kernel rule is applied only once while multi-rules are applied as often as suitable matches are found. For formal details on amalgamation and a large case study, see [8].

Figure 6 shows the interaction scheme with a kernel rule and two multi-rules specifying the firing of a Petri net transition with arbitrary many pre- and post-domain places. The kernel rule has the same application condition as rule ActivationCheck (see Figure 1). Note that we do not need the isActivated flag any more because the check and the complete firing step is performed by a single application of the amalgamated rule. The amalgamated rule constructed for example along a match to an activated transition with three pre- and two post-domain places is similar to the rule in Figure 5 but has no NACs because the match into the host graph is predefined by construction.

Construction of Inverse Rules. For a given rule, the inverse rule has LHS and RHS exchanged. Moreover, application conditions are shifted over the rule. Figure 7 shows the inverse rule of rule AddPost (see Figure 2), where an existing token is removed. The shifted NAC requires that there is exactly one token on the place for the rule to be applicable.

Construction of Minimal Rules. A new challenge from MDD comes from the field of *model versioning*, where the new notion of *graph modification* [14], a span $G \leftarrow D \rightarrow H$ has been established to formalize model differences for visual models. Based on graph modifications, so-called *minimal rules* may be extracted from a given span to exploit conflict detection techniques for rules. A minimal rule comprises the effects of a given rule in a minimal context. Via context menu, AGG 2.0 supports the extraction of a minimal rule from a selected rule (interpreted as graph modification). For example, the minimal rule of the rule in Figure 5 does not contain the arcs connecting the places and transitions, since

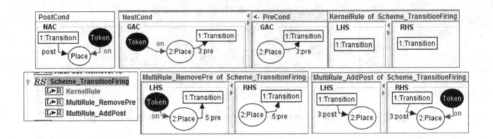

Fig. 6. Interaction scheme defining a transition firing step in Petri nets

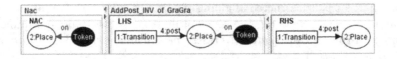

Fig. 7. Inverse rule of Rule AddPost

these arcs are not changed by the rule. It contains the place nodes (because edges connected to them are deleted or generated), the token nodes (either deleted or generated) and the transition node (its attribute is changed).

4   Tuning the Critical Pair Analysis

The critical pair analysis (CPA) checks rule pairs for conflicts of rule applications in a minimal context. An example for a conflict in a minimal context is shown in Figure 8, where the classical *forward conflict* in Petri nets is detected when analyzing the rule pair (RemovePre, RemovePre). As indicated in the conflict view, we have a *delete-use conflict* since two transitions need to remove the same token from their common pre-domain place.

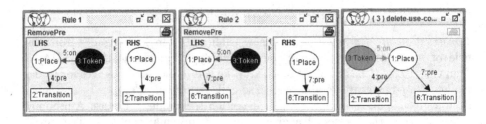

Fig. 8. CPA of rule pair (RemovePre, RemovePre), detecting a forward conflict

Selection of rules for CPA. So far, the CPA could be evoked on a graph grammar, yielding the critical pairs for each possible rule pair of the grammar. AGG 2.0 provides free selection of rule sets to be analyzed. This feature has shown to be very convenient, e.g., for a case study, where self-healing systems are modeled by several rule sets (normal system behaviour rules, context-changing rules, repair rules). These sets are analyzed for conflicts with each other [6].

Modularization of a model into different sub-grammars. In AGG 2.0, rule sets may be imported into an existing grammar (provided the type graph of the imported grammar is a subgraph of the type graph of the importing grammar). This supports modularization of a model without destroying the possibility to analyze the complete system.

Interrupt and resume running CPA. This usability feature is very handy for complex computations which may take some time. A partial CPA result may be stored and reloaded in AGG 2.0.

Generation of Filter NACs. If during CPA, critical pairs are found that are analyzed by the user as not causing real conflicts, additional NACs for these rules may be generated automatically that contain the critical overlapping region. A new CPA of the rules together with the new *Filter NACs* does not show the previous critical pair anymore [10].

5   Related Work and Conclusions

AGG is one of the standard graph transformation tools implementing the algebraic approach as presented in [5]. Other graph transformation tools, such as Fujaba [7], ViaTra [4], VMTS [3], GrGen [16], and Groove [17], implement different kinds of graph transformation approaches. Some kinds of rule application control structures are offered by all of these tools, e.g., Fujaba uses story diagrams, a kind of activity diagrams. Groove also supports nested application conditions as well as universal quantification using amalgamation.

Concerning the verification of graph transformation systems, VIATRA and Groove concentrate on some kind of model checking, while AGG is the only tool that consequently implements the theoretical results available for algebraic graph transformation. These results are mainly concerned with conflict and dependency detection of rules and static applicability checks for rule sequences. The analysis features offered by AGG are also used by our graph transformation-based tools TIGER (a tool for generating visual editor and simulator plug-ins in ECLIPSE [2]) and HENSHIN (an EMF model transformation engine [1]).

AGG 2.0 extends the existing features now coherently with support for application conditions and object flow, and for automatic construction of amalgamated, concurrent, inverse and minimal rules. Moreover, the critical pair analysis has become more usable due to experiences made in several case studies.

References

1. Arendt, T., Biermann, E., Jurack, S., Krause, C., Taentzer, G.: Henshin: Advanced Concepts and Tools for In-Place EMF Model Transformations. In: Petriu, D.C., Rouquette, N., Haugen, Ø. (eds.) MODELS 2010, Part I. LNCS, vol. 6394, pp. 121–135. Springer, Heidelberg (2010)
2. Biermann, E., Ehrig, K., Ermel, C., Hurrelmann, J.: Generation of simulation views for domain specific modeling languages based on the Eclipse Modeling Framework. In: Automated Software Engineering (ASE 2009), pp. 625–629. IEEE Press (2009)
3. Budapest University of Technology and Economics: Visual Modeling and Transformation System (VMTS) (2010), http://www.aut.bme.hu/Portal/Vmts.aspx
4. Eclipse Consortium: VIATRA2 (Visual Automated Model Transformations) Framework (2011), http://www.eclipse.org/gmt/VIATRA2/
5. Ehrig, H., Ehrig, K., Prange, U., Taentzer, G.: Fundamentals of Algebraic Graph Transformation. EATCS Monographs in Theoretical Computer Science. Springer (2006)
6. Ehrig, H., Ermel, C., Runge, O., Bucchiarone, A., Pelliccione, P.: Formal Analysis and Verification of Self-Healing Systems. In: Rosenblum, D.S., Taentzer, G. (eds.) FASE 2010. LNCS, vol. 6013, pp. 139–153. Springer, Heidelberg (2010)
7. Fujaba Development Group: Fujaba Tool Suite (2011), http://www.fujaba.de/
8. Golas, U., Biermann, E., Ehrig, H., Ermel, C.: A visual interpreter semantics for statecharts based on amalgamated graph transformation. In: Echahed, R., Habel, A., Mosbah, M. (eds.) Selected Papers of International Workshop on Graph Computation Models (GCM 2010). ECEASST, vol. 39 (2011)
9. Habel, A., Pennemann, K.H.: Correctness of high-level transformation systems relative to nested conditions. Mathematical Structures in Computer Science 19, 1–52 (2009)
10. Hermann, F., Ehrig, H., Golas, U., Orejas, F.: Efficient analysis and execution of correct and complete model transformations based on triple graph grammars. In: Proc. Int. Workshop on Model Driven Interoperability, pp. 22–31. ACM (2010)
11. Jurack, S., Lambers, L., Mehner, K., Taentzer, G., Wierse, G.: Object Flow Definition for Refined Activity Diagrams. In: Chechik, M., Wirsing, M. (eds.) FASE 2009. LNCS, vol. 5503, pp. 49–63. Springer, Heidelberg (2009)
12. Lambers, L.: Certifying Rule-Based Models using Graph Transformation. Ph.D. thesis, Technische Universität Berlin (2009)
13. Taentzer, G.: AGG: A Graph Transformation Environment for Modeling and Validation of Software. In: Pfaltz, J.L., Nagl, M., Böhlen, B. (eds.) AGTIVE 2003. LNCS, vol. 3062, pp. 446–453. Springer, Heidelberg (2004)
14. Taentzer, G., Ermel, C., Langer, P., Wimmer, M.: Conflict Detection for Model Versioning Based on Graph Modifications. In: Ehrig, H., Rensink, A., Rozenberg, G., Schürr, A. (eds.) ICGT 2010. LNCS, vol. 6372, pp. 171–186. Springer, Heidelberg (2010)
15. TFS-Group, TU Berlin: AGG 2.0 (2011), http://tfs.cs.tu-berlin.de/agg
16. Universität Karlsruhe: Graph Rewrite Generator, GrGen (2010), http://www.info.uni-karlsruhe.de/software.php/id=7
17. University of Twente: Graphs for Object-Oriented Verification (GROOVE) (2011), http://groove.cs.utwente.nl/

Integration of a Pattern-Based Layout Engine into Diagram Editors

Sonja Maier and Mark Minas

Universität der Bundeswehr München, Germany
{sonja.maier,mark.minas}@unibw.de

Abstract. In this paper, we outline our pattern-based layout approach and its integration into a diagram editor. In particular, we summarize editor features that were made possible by the approach. Each layout pattern encapsulates certain layout behavior. Several layout patterns may be applied to a diagram simultaneously, even to overlapping diagram parts. Our approach includes a control algorithm that automatically deals with such situations. To support the user in an interactive environment, it is not sufficient to apply the same layout behavior in every situation. Instead, the user also wants to select and alter the layout behavior at runtime. Our approach as well as the editor features described in this paper are specifically designed for such an environment.

1 Introduction

A layout engine usually runs continuously within diagram editors and improves the layout in response to user interaction in real-time. Layout improvement includes all sorts of changes concerning the position or shape of diagram components. For instance in Figure 1, if class C is moved, the end point of the connected association is updated accordingly. It is, however, not reasonable to completely automate layout improvements in diagram editors. The editor user would also like to influence the layout at runtime.

We have developed a *pattern-based layout approach* that is tailored to such an environment [2]. A layout pattern, which encapsulates certain layout behavior, consists of a (*pattern-specific*) *meta model* together with some *assertions* and a *layout algorithm*, which are defined using this meta-model. A pattern may be applied to a diagram, which means that a pattern instance is created and bound to (parts of) the diagram. The assertions "define" a valid layout of the diagram (part), and the layout algorithm is responsible for "repairing" the layout when assertions are violated. The diagram layout is affected by this pattern instance until it is deleted. A diagram layout is usually defined by several instances of layout patterns whose layout algorithms cannot be executed independently. A *control algorithm* automatically coordinates when and how these layout algorithms are executed [3].

In this paper, we give an overview of the layout approach with the help of a class diagram editor as a simple example.[1] We focus on some editor features that support the user during the process of layout improvement:

---

[1] A screencast is available at www.unibw.de/inf2/DiaGen/Layout.

A. Schürr, D. Varró, and G. Varró (Eds.): AGTIVE 2011, LNCS 7233, pp. 89–96, 2012.

User-controlled and Automatic Layout: The editor *user* can apply layout patterns for user-selected parts of the diagram. These pattern instances affect the diagram layout until the user deletes them. Furthermore, the specification of the diagram editor describes when layout patterns are *automatically* applied depending on the diagram's syntactic structure.

Pattern Instance Visualization: Pattern instances created by the editor user are visualized in the diagram.

Syntax Preservation: The editor makes sure that the layout engine does not modify the syntactic structure of the diagram.

Layout Suggestions: The editor can automatically suggest application of layout patterns when the user selects some part of the diagram.

Automatic as well as user-controlled layout and pattern instance visualization as well as syntax preservation have been described or at least sketched earlier [2,3], but layout suggestions have not been published yet. The concepts described in this paper have been completely implemented and examined in several diagram editors.

The term *layout pattern* has been coined by Schmidt et al. in the context of syntax-directed editors [5]. They use tree grammars for specifying the digram language syntax whereas we use meta-models, which are more widely accepted than grammars. In their approach, layout is computed via attribute evaluation whereas we support several ways of layout computation, in particular constraint solving techniques and off-the-shelf graph drawing algorithms. There also exist some editors including a layout engine that are somewhat related to our approach. One of these editors is DUNNART [1], a graph editor, which is based on declarative constraints and which provides some user-controlled layout behavior. Most of these tools have in common that the layout engine is hand-coded and hard-wired in the editor. Our approach supports modular specification of layout on an abstract level and using this layout behavior in different editors.

Section 2 introduces the class diagram editor as a running example and describes its layout capabilities. Our layout approach is sketched in Section 3. Section 4 outlines the integration of the approach into an editor. Section 5 concludes the paper.

2   Running Example

Figure 1 shows a class diagram editor, which has been created with the editor generation framework DIAMETA [4]. DIAMETA allows for generating visual language editors from specifications of their visual languages. The core of the specification is a meta-model (called *language-specific meta model*, LMM, in the following) of the diagram language, here class diagrams. The meta-model describes the language's abstract as well as concrete syntax.

Several layout patterns have been integrated into the class diagram editor: **Non-Overlap** removes overlapping of packages and classes that are (directly)

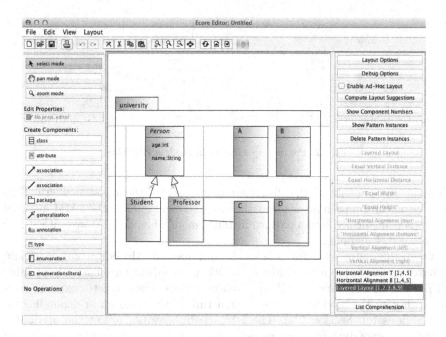

Fig. 1. Running example: Class diagram editor

contained in the same package by applying a force-directed layout algorithm.
Containment ensures the correct nesting of packages and classes. **List** arranges
attributes of classes as a list. **Minimal Size** enforces a minimal size of packages
and classes. **Edge Connector** ensures that associations and generalizations stay
attached to classes. **Alignment** (horizontal & vertical) aligns certain packages
and classes vertically or horizontally respectively. **Equal Distance** (horizontal
& vertical) makes sure that certain packages and classes have an equal distance
to each other. **Equal Size** (width & height) makes sure that certain packages
and classes have the same height or width respectively. Finally, **Layered Layout**
assigns each class to a certain horizontal layer such that generalization arrows
always point upward.

3   Layout Engine

Each layout pattern is defined on a pattern-specific meta-model (PMM) whereas
the diagram language is specified using a language-specific meta-model (LMM)
as outlined in Sections 1 and 2. Figure 2 shows three PMM examples: The Containment PMM used by the Containment pattern, List PMM used for the List
pattern, and Graph PMM used for the Layered Layout pattern. The PMMs are
apparently independent of a concrete diagram language and its meta-model. In
order to use a layout pattern for a certain diagram language, the language's
LMM must be mapped to the pattern's PMM. This mapping identifies the roles

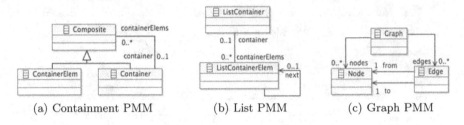

(a) Containment PMM   (b) List PMM   (c) Graph PMM

Fig. 2. Pattern-specific meta-models (PMMs)

that the components of the visual language play in the layout pattern. Pattern instantiation then means that the diagram's model, i.e., the LMM instance (LM), is transformed into a pattern instance model, i.e., a PMM instance (PM), following this mapping.

When the editor user draws or modifies a diagram, the editor automatically updates its diagram model (LM). In addition, the editor automatically updates all pattern instance models (PMs), creates new ones and deletes existing ones depending on the modified syntactic structure of the diagram or depending on user actions (cf. Section 4.1). The PM attribute values reflect the attributes of the corresponding diagram model. The pattern's assertions (cf. Section 1) are checked on these attribute values. The layout algorithm modifies the PM attribute values when one or more assertions are violated. These values are then transformed back to the corresponding LM attribute values following the mapping between LMM and PMM. The layout is updated this way.

Of course, the layout algorithms must not modify the PM attributes and, therefore, the diagram model attributes independently. The modifications must rather be coordinated. This is the task of the control algorithm, which was described in detail in [3]. Via backtracking, this algorithm determines the order in which the layout algorithms of the pattern instances change the diagram. This must be done at runtime because the layout can only be computed as soon as all attribute values – including the attribute values the user changed – are known. The idea of the control algorithm is that changes made by the user are propagated through the diagram, trying to find a layout satisfying all assertions. Starting with the component(s) (or more precisely, attribute(s)) changed by the user, all pattern instances are checked that involve these components. Each pattern instance whose assertions are violated changes one or more attribute value(s) by its encapsulated layout algorithm in order to "repair" it. Afterwards, all further pattern instances that involve these attributes are checked. This procedure is continued until all pattern instances are satisfied, or until the algorithm signals a failure. A failure is signaled if there does not exist a "valid" layout, which might be the case if two or more pattern instances contradict each other. The algorithm also signals a failure in case layout computation would take longer than a user-configurable threshold.

With the help of layout patterns, different layout algorithms can be combined. The following three types of layout algorithms are currently supported:

- **Standard graph drawing algorithms** may be used, e.g., the Sugiyama algorithm for the creation of a layered layout. These algorithms can be used off-the-shelf from graph drawing libraries.
- **Declarative constraints** may be used to specify the layout algorithm which then uses a constraint solver for computing the layout. With this type of layout algorithm, global layout behavior such as "several classes have an equal distance to each other" can be defined.
- In addition, our own version of **rule-based layout algorithms** may be used, which is specifically tailored to the interactive nature of visual language editors. The rule-based layout algorithm consists of a set of assertions that "define" the layout, and a set of rules that "repair" violated assertions and hence update the layout after user modifications.

It is possible to define quite complex layout behavior with these types of layout algorithms. However, layout patterns should be chosen carefully because the diagram layout is computed and updated at runtime. Especially using a constraint solver may lead to a bad performance, e.g., if it aims at minimizing edge crossings.

4   Integration of the Layout Engine into an Editor

We specified several layout patterns, and integrated them into various visual language editors. For instance, they were integrated into the class diagram editor presented earlier, into a graph editor, into a GUI forms editor that allows the user to create GUIs, or into a VEX editor that allows the user to draw lambda expressions visually.

The creation of the layout modules as well as their inclusion in different visual language editors turned out to be (more or less) straightforward. Furthermore, the concept of layout patterns allowed us to develop some features that increased the usability of the editors. Some of these features will be described in the following.

4.1   Automatic and User-Controlled Layout

A layout pattern may be applied to a diagram if some part of the diagram model can be transformed to a pattern instance model. This situation must be specified when generating the editor. This is done as follows: The generated editor uses a graph (actually a hypergraph) for representing the diagram model. This hypergraph corresponds to the language-specific meta-model (LMM). Pattern instance models (PMs) are internally represented by graphs, too. Transformations of the diagram model to the PMs are realized by graph transformations. The situation that a layout pattern may be applied to a diagram, therefore, is simply the situation that such a graph transformation may be applied.

The layout engine supports two modes of layout pattern instantiation: *automatic and user-controlled application* of layout patterns. **Automatic instantiation** means that the diagram editor tries to transform the diagram model to pattern instance models, i.e., to apply the graph transformations whenever possible. This mode of pattern instantiation makes sense for layout aspects that are essential for a diagram language. In the class diagram editor, these are the application of the non-overlap pattern, containment pattern, list pattern, minimal size pattern, and edge connector pattern to the corresponding diagram parts.

However, automatic instantiation does not make sense in situations where the user would like to modify the layout for, e.g., mere aesthetic reasons. In the class diagram editor, these are, e.g., the horizontal alignment of classes. Applying such a layout pattern to every possible situation does not make sense. The editor user rather has to indicate the situations where he wishes to apply a pattern. This is realized by **user-controlled instantiation**. The editor specification must contain graph transformation rules (actually graph transformation programs in DIAMETA) for a layout pattern. By selecting a layout pattern and one or more components, the editor user chooses a part of a diagram for which the layout pattern shall be instantiated. The selected diagram components correspond to certain graph components of the diagram model. The editor then tries to apply a graph transformation specified for the selected layout pattern to the diagram model that corresponds to the selected graph components of the diagram model. After creation, the pattern instance is updated after each diagram modification until it is deleted, either by the editor user who explicitly deletes this pattern instance, or when the diagram is modified in a way such that the layout pattern does no longer fit. In the class diagram editor, the equal distance pattern, the alignment pattern, the equal size pattern and the layered layout pattern are controlled by the user. All pattern instances that were created by the user are shown in a list at the bottom-right of the editor.

In the example shown in Figure 1, the user has created the following pattern instances: Classes **Person**, **Student** and **Professor** and the corresponding generalizations are rearranged by a layered layout algorithm. Classes **Person**, **A** and **B** are aligned horizontally at the top *and* at the bottom. Some pattern instances have been automatically created. When the user modifies the diagram, the two alignment pattern instances and the layered layout pattern instance preserve the layout that was chosen by the user. The non-overlap pattern instance moves the classes that are contained in the package **university** to assure that they do not overlap. The containment pattern instance moves the package and the classes to preserve the correct nesting of the package **university** and its contained classes. The list pattern instance correctly arranges the attributes in the class **Person**. The minimal size pattern instance enforces a minimal size of the package and the classes. Finally, the edge connector pattern instance updates the start and end point of the association and the generalizations to keep them correctly connected to the classes.

4.2   Pattern Instance Visualization

Experiments have shown that users of diagram editors have trouble "understanding" the layout dependencies between components if current layout pattern instances are not displayed. Therefore, diagram editors display currently active layout pattern instances in the diagram. In the example shown, instances of the horizontal alignment pattern are visualized via colored lines, and instances of the layered layout pattern via colored boxes. It is possible to create instances of the same layout pattern several times in one diagram. To distinguish these pattern instances, they are highlighted in different colors.

If the user selects one of the pattern instances in the list at the bottom-right of the editor, the involved components are highlighted via a gray cross in the middle of each component. For instance, in the editor shown in Figure 1, the user has selected the layered layout pattern instance, and hence the classes Person, Student and Professor and the two generalizations are highlighted.

4.3 Syntax Preservation

Freehand editors allow to arrange diagram components on the screen without any restrictions, and the syntax is defined by the location of the components on the screen. Because a layout pattern instance does not know about the diagram syntax, it can easily rearrange components such that the diagram's syntactic structure is modified. This is usually unwanted behavior, and it should be possible to prevent it. For instance, in the class diagram editor, a syntax change may occur if the layout engine moves one class on top of another class (this is possible only if the non-overlap pattern is turned off) that has an association connected to it. Then it is not clear to which class this association is connected.

Syntax preservation has been added for DIAMETA editors. It is achieved by comparing the internal graph representation of the diagram model before and after a layout modification. If the graph structure is changed by the layout engine, the user changes that led to this violation are undone.

4.4 Layout Suggestions

To further support the editor user, the editor is able to suggest layout patterns that could be applied to a user-selected part of the diagram. Layout suggestions are computed by "trying out" each layout pattern. For each layout pattern, pattern matching is performed in the user-selected part of the diagram. For the maximal match found, a pattern instance is created. The layout engine then tries to find a valid solution, taking into account the newly created pattern instance(s) and all other pattern instances currently present in the diagram. The pattern is applicable if the layout engine is able to compute a valid layout. In addition, the attribute changes that are performed by the layout engine are examined. A metric is used to rate the applicability of the different layout patterns. Layout patterns that result in minor diagram changes are favored over layout patterns that result in major diagram changes.

This feature can be used as follows: After selecting one or more components, the user can click the button *Compute Layout Suggestions*. The buttons on the right side of the editor are then highlighted in a certain color. Gray indicates that the corresponding pattern cannot be applied because it either does not fit the chosen diagram part or is inconsistent with the currently active pattern instances. Blue indicates that the corresponding pattern can be applied. Asterisks are added to the label if the application of the pattern results in minor changes of the diagram, and they are omitted if it results in major changes.

In Figure 1, layout suggestions for classes `Professor`, `C` and `D` were computed. Button *Horizontal Alignment (top)*, e.g., is highlighted in blue, and asterisks are added as only minor diagram changes would be necessary after applying the pattern, whereas button *Vertical Alignment (left)* is highlighted in blue, and no asterisks are added as major diagram changes would be necessary.

5   Conclusions

We have sketched our pattern-based layout approach and the integration of the layout engine into a diagram editor. We focused on some features that are made possible by the pattern-based layout approach, namely user-controlled layout, pattern instance visualization, syntax preservation and layout suggestions. We have further features in mind that are enabled by the pattern-based approach and that could also be included, e.g., automatic layout after diagram import.

Our pattern-based approach as well as the features described have been completely integrated into DIAMETA. Furthermore, we created several visual language editors that include the layout engine, and we observed that the layout engine produces good results, and that the overall performance is satisfactory.

During the development of the layout approach and its integration into several editors, some questions arose: Which layout patterns should be included in a certain visual language editor? How should a layout pattern be visualized? As a next step, we plan to perform a user study to answer these questions.

References

1. Dwyer, T., Marriott, K., Wybrow, M.: Dunnart: A Constraint-Based Network Diagram Authoring Tool. In: Tollis, I.G., Patrignani, M. (eds.) GD 2008. LNCS, vol. 5417, pp. 420–431. Springer, Heidelberg (2009)
2. Maier, S., Minas, M.: Pattern-based layout specifications for visual language editors. In: Proc. of the 1st International Workshop on Visual Formalisms for Patterns. ECEASST, vol. 25 (2009)
3. Maier, S., Minas, M.: Combination of different layout approaches. In: Proc. of the 2nd International Workshop on Visual Formalisms for Patterns. ECEASST, vol. 31 (2010)
4. Minas, M.: Generating meta-model-based freehand editors. In: Proc. of the 3rd International Workshop on Graph Based Tools. ECEASST, vol. 1 (2006)
5. Schmidt, C., Kastens, U.: Implementation of visual languages using pattern-based specifications. Software: Practice and Experience 33(15), 1471–1505 (2003)

Tool Demonstration of the Transformation Judge

Steffen Mazanek, Christian Rutetzki, and Mark Minas

Universität der Bundeswehr, München, Germany
steffen.mazanek@gmail.com, christian.rutetzki@paprots.com,
mark.minas@unibw.de
http://www.unibw.de/inf2/

Abstract. The *transformation judge* is a novel system for the automatic evaluation and comparison of graph and model transformations that have been submitted as solutions for common transformation tasks such as those accepted as case studies for the transformation tool contest. The most important feature of this system is the correctness check that is done by black-box-testing. But also performance data and other information about the solutions are collected. So, for academic as well as industrial users of transformation tools, the judge could be a good starting point for choosing a particular transformation tool for their respective task, since they can easily explore and compare different solutions for similar tasks.

In this demonstration we show the most important use cases of the judge, i.e., uploading of cases and corresponding solutions as well as the automatic evaluation and comparison of solutions.

1 Introduction

We have almost completed our work on the so-called transformation judge,[1] which is an online judge system for graph and model transformations, i.e., it allows to automatically evaluate solutions for common transformation tasks. The idea of an online judge originally stems from the domain of programming contests, where such systems are used to give immediate feedback about the correctness of submitted solutions. In this context, however, correctness only means that a certain set of reference inputs is transformed by the submitted program into the corresponding reference outputs, i.e., black-box-tests are performed. Both inputs as well as outputs are given as plain text in this domain and the actual output can be compared with the reference output character by character; with graphs/models as the main artefacts this step certainly is more complicated. Several online judge systems are available that, among others, differ in the supported set of programming languages. Most widely known is the UVa Online Judge,[2] which is used in the well-known ACM programming contests,[3] but also by people who just want to train and improve their programming skills.

---

[1] http://sites.google.com/site/transformationjudge/
[2] http://uva.onlinejudge.org/
[3] http://icpc.baylor.edu/

A. Schürr, D. Varró, and G. Varró (Eds.): AGTIVE 2011, LNCS 7233, pp. 97–104, 2012.
© Springer-Verlag Berlin Heidelberg 2012

From the organization of the transformation tool contest TTC[4] the idea was born to provide an online judge for the graph and model transformation community. In previous editions of the contest the solution submitters presented their solutions for the respective transformation cases in front of a critical audience that acted as a jury and ranked the presented approaches according to several criteria. Of course, such a voting is highly inappropriate for criteria such as correctness, but there was no other option at that time. Actually, correctness was even among the most subjective criteria as indicated by a very high variance in the votings. The transformation judge solves this problem by carrying over the online judge concept from programming contests. Moreover, the transformation judge allows to compare transformation approaches with respect to further criteria such as performance, lines of code or user rating and, even more importantly, across various cases. So, for both academic and industrial users of transformation tools, the judge, once populated with cases and solutions, could be a good starting point for choosing a particular transformation tool for their respective task, since they can easily explore and compare different solutions for similar tasks. Finally, the transformation judge will be a platform for beginners, who want to get an impression of the different transformation approaches. To simplify this learning step a Hello World case [7] that comprises very basic transformation tasks was selected as one of the cases for TTC 2011 [11] and resulted in an initial fill of the judge.

There are several challenges when developing a transformation judge compared to a conventional judge for programming algorithms, most importantly:

- Many transformation tools are still at the prototypical level, in contrast to the programming languages supported by the conventional judges.
- The different transformation approaches deal with varying input and output formats.
- Several transformation approaches are integrated into complex graphical tools and difficult to invoke in headless mode.
- Regarding security, one needs to be not only careful with the actual transformations but also with the zoo of transformation tools to be integrated into the judge.
- For most cases, the output of a transformation cannot be compared on a per-character-basis with the reference output as in conventional judges, because graphs/models have different possible, equivalent string representations; even worse, in general it is inherently expensive to compare two graphs/models of reasonable sizes in practical time.

This demonstration paper describes the most important use cases of the judge, i.e., the upload of cases and corresponding solutions as well as the automatic evaluation and comparison of solutions. For a better understanding, also the architecture of our system is briefly introduced.

---

[4] http://planet-research20.org/ttc2011/

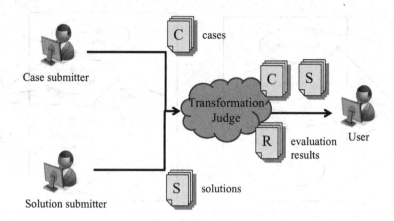

Fig. 1. The main roles of judge users

2 Use Cases of the Transformation Judge

The main roles of judge users and the basic process behind the judge are shown in Figure 1. Case submitters upload cases, which basically consist of a task description, an input metamodel, an output metamodel and a set of pairs of reference input models with their corresponding output models. They can also set some more parameters such as the maximal execution time after which a solution is terminated. After the publication of the case, solution submitters upload their respective solutions. To this end, the actual transformation code needs to be uploaded and the used transformation language needs to be selected from a list of supported transformation languages. Besides the case and solution submitters there are also normal users who benefit from the system by exploring and comparing the various evaluation results.

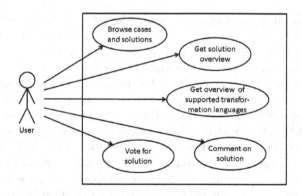

Fig. 2. Use cases of a normal user

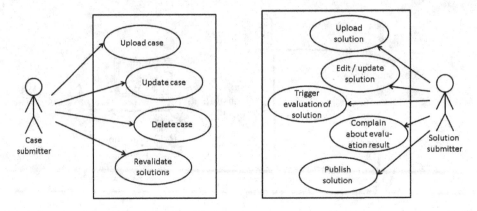

Fig. 3. Use cases of case and solution submitters

Figure 2 shows the system's use cases from the perspective of a normal user. More precisely we distinguish anonymous users from registered users, and certain functions such as commenting or voting on solutions can only be used by registered users.

Figure 3 shows the system's use cases from the perspective of case and solution submitters. A case, i.e., a transformation task, consists of a name, a description file, the determined model compare procedure and a set of model pairs (reference inputs and expected outputs) that are used for black-box testing the solutions. Since a case, and thus the model pairs, can be changed during the update of the case, there need to be mechanisms for versioning and also for checking whether the already uploaded solutions still pass the black-box test. This mechanism can be triggered by the case submitter and is called "Revalidate solutions" in the figure. Solutions that still comply to the new version of the case are automatically connected with this new version.

3   System Architecture

The judging system is implemented in a modular way following a service-oriented approach. All supported transformation tools and also the comparison tool are addressed as webservices. Figure 4 shows how the transformation judge actually evaluates a solution. After the upload of a solution to the judge website its evaluation can be triggered. First, the required case- and solution-related data need to be gathered from the database. It is passed to the evaluation manager that is responsible for controlling the computation of an evaluation result. To this end, first the transformation tool responsible for the specific solution needs to be invoked, which is done via its webservice. The result that is passed back consists of logging information, performance data, and, most importantly, the output model. This model needs to be compared with the expected output, i.e., the reference output model. This job is performed by the comparator webservice, which returns some logging information including a human readable

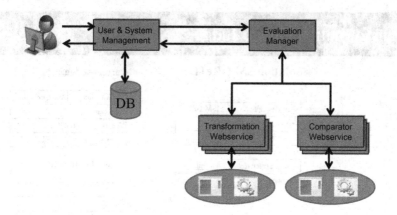

Fig. 4. The steps for the evaluation of a solution

description of the comparison result and, if applicable, a formal difference model. The transformation and compare step need to be repeated for each pair of reference input/output models. From all these results an overall evaluation result is constructed, passed back and shown to the user, who can publish his solution afterwards. If he is not satisfied with the evaluation result he can improve his solution until it passes the test. Note that all communication between the webservices is performed asynchronously.

In order to integrate a transformation tool, the tool developers need to contribute their tool as an executable file that has to follow a particular call schema. GReTL [2], ETL [5] and plain Java are already supported and ATL [4], VIATRA [1] and PETE [10] will follow soon. The judge then wraps these executables into webservices.

Restrictions

The judge currently has the following restrictions:

- The judge (for now) only provides support for EMF models, i.e., input and output metamodels have to conform to EMF's Ecore metamodel.[5] Note that EMF is a de facto modeling standard nowadays and many transformation tools are based on EMF or at least provide EMF import/export facilities.
- The judge (for now) only provides support for transformation tasks that require exactly one input and one output model and metamodel, respectively.
- The whole transformation currently has to be uploaded as a single file.

Focusing on the EMF format also has allowed us to integrate model comparison in a quite straightforward way, because there are several tools available that can compare EMF models. EMFCompare[6] [9] probably is the most widely used tool

---

[5] http://www.eclipse.org/emf/
[6] http://www.eclipse.org/emf/compare/

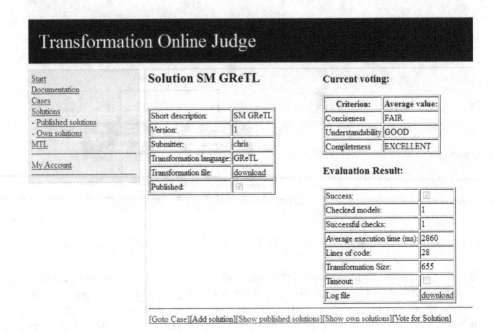

Fig. 5. Screenshot of the evaluation screen

for this purpose and, thus, we integrated this tool. Our experiences regarding the use of model comparison tools for our purpose and also the issues we had to deal with are described in detail in [8].

4   Related Work

Besides the already mentioned UVa judge, there are also several smaller systems such as the sphere judge [6] or the z-training judge,[7] which support different programming languages or provide further features. The latter, for instance, provides interactive graders as a means for the evaluation of intermediate results of a computation. Such a feature might be interesting for the evaluation of very complex or step-wise transformations. We have decided to implement our transformation judge from scratch instead of extending an existing system, because discussions with both the developers of the sphere and the z-training judge revealed that it would have been a high effort to adapt these systems to support transformation tasks. Also, a modular architecture with tool-specific webservices was an important argument for a completely new system.

Further, there are several approaches that address the evaluation and comparison of transformations, among them the yearly tool contest. One of the most important contributions, however, is the benchmark proposed by Varró et al. as

---

[7] http://www.z-training.net/

a means for the systematic comparison of the performance of transformation tools [12]. Setting up such a benchmark from scratch is a huge effort, because the different systems ideally should be installed in the same environment or at least on comparable hardware. The transformation judge can be used for benchmarking as a side effect, because performance data is collected automatically while running the transformations.

5   Conclusion

We have only put a few quite basic transformation tasks into the judge so far. The judge deals with them easily and provides quite readable information. It has successfully verified the GReTL solutions [3] to some subtasks of this year's Hello-World case. Note that GReTL is a graph transformation tool. This shows that the judge is not only dedicated to model transformation tools (although many graph transformation tools do still not support EMF).

At the moment the webservices for both the transformation and the compare tools can only be accessed by the judge server itself. However, it would be possible to make them publicly available. This would, e.g., enable users to compare models in a very lightweight way and to invoke different transformation tools. So, users would not need to install tools locally in order to evaluate their suitability for the task at hand.

Note that the transformation judge is not online yet. The described functionality is mostly implemented, but there are a few security issues that still need to be addressed. A screenshot of the web interface is shown in Figure 5. Moreover, screencasts of the system are available at the project website: `http://sites. google.com/site/transformationjudge/`.

Due to the challenges listed in this paper, the transformation judge is interesting from a software engineering perspective. However, its real value hopefully will be revealed in the future when the different transformation approaches can be compared in an objective way across a reasonable number of cases.

Plenty of work remains to be done in the future. For instance, it would be good to divide the test cases into public and private ones, the further for debugging solutions and the latter for the real competition, i.e., without exposing the models. Also, we need to compute a better indicator for conciseness than just lines of code. And, finally, we have to address the current restrictions of the system to make it more widely applicable.

Acknowledgements. A great thank you to the transformation tool developers that helped with the integration of their respective tools (Tassilo Horn, Louis Rose, Massimo Tisi). Also many thanks to Aleksandar Zlateski, the developer of z-training judge, for a lot of advice regarding the implementation of an online judge and Cédric Brun and Patrick Konemann for EMFCompare tips. Finally, a special thank you to Pieter van Gorp, who discussed this project with us from the very beginning.

References

1. Csertán, G., Huszerl, G., Majzik, I., Pap, Z., Pataricza, A., Varró, D.: VIATRA –
   Visual automated transformations for formal verification and validation of UML
   models. In: 17th IEEE International Conference on Automated Software Engineering (ASE 2002), pp. 267–270. IEEE Computer Society (2002)
2. Ebert, J.: Metamodels taken seriously: The TGraph approach. In: Proc. of the 12th
   European Conference on Software Maintenance and Reengineering, CSMR 2008,
   p. 2. IEEE (April 2008)
3. Horn, T.: Saying Hello World with GReTL – A solution to the TTC 2011, instructive case. In: Van Gorp et al. [11]
4. Jouault, F., Allilaire, F., Bézivin, J., Kurtev, I.: ATL: A model transformation
   tool. Science of Computer Programming 72(1-2), 31–39 (2008)
5. Kolovos, D.S., Paige, R.F., Polack, F.A.C.: The Epsilon Transformation Language.
   In: Vallecillo, A., Gray, J., Pierantonio, A. (eds.) ICMT 2008. LNCS, vol. 5063, pp.
   46–60. Springer, Heidelberg (2008)
6. Kosowski, A., Małafiejski, M., Noiński, T.: Application of an Online Judge & Con-
   tester System in Academic Tuition. In: Leung, H., Li, F., Lau, R., Li, Q. (eds.)
   ICWL 2007. LNCS, vol. 4823, pp. 343–354. Springer, Heidelberg (2008)
7. Mazanek, S.: Hello world! An instructive case for the Transformation Tool Contest.
   In: Van Gorp et al. [11], http://sites.google.com/site/helloworldcase/
8. Mazanek, S., Rutetzki, C.: On the importance of model comparison tools for the
   automatic evaluation of the correctness of model transformations. In: Proceedings
   of the 2nd International Workshop on Model Comparison in Practice, pp. 12–15.
   ACM (2011)
9. Mülder, A., Schill, H., Wendehals, L.: Modellvergleich mit EMF Compare – Teil 1:
   Funktionsweise des Frameworks. Eclipse Magazin 4, 43–47 (2009)
10. Schätz, B.: Formalization and Rule-Based Transformation of EMF Ecore-Based
    Models. In: Gašević, D., Lämmel, R., Van Wyk, E. (eds.) SLE 2008. LNCS,
    vol. 5452, pp. 227–244. Springer, Heidelberg (2009)
11. Van Gorp, P., Mazanek, S., Rose, L. (eds.): Proc. of the Fifth Transformation
    Tool Contest, Zürich, Switzerland. Electronic Proceedings in Theoretical Computer
    Science, vol. 74 (2011)
12. Varró, G., Schürr, A., Varró, D.: Benchmarking for Graph Transformation. In:
    Erwig, M., Schürr, A. (eds.) Proc. IEEE Symposium on Visual Languages, pp.
    79–100. IEEE Computer Society Press (2005)

Knowledge-Based Graph Exploration Analysis

Ismênia Galvão[1], Eduardo Zambon[2,*], Arend Rensink[2],
Lesley Wevers[2], and Mehmet Aksit[1]

[1] Software Engineering Group,
{i.galvao,m.aksit}@ewi.utwente.nl
[2] Formal Methods and Tools Group,
Computer Science Department,
University of Twente
PO Box 217, 7500 AE, Enschede, The Netherlands
{zambon,rensink}@cs.utwente.nl, l.wevers@student.utwente.nl

Abstract. In a context where graph transformation is used to explore
a space of possible solutions to a given problem, it is almost always necessary to inspect candidate solutions for relevant properties. This means
that there is a need for a flexible mechanism to query not only graphs
but also their evolution. In this paper we show how to use Prolog queries
to analyse graph exploration. Queries can operate both on the level of
individual graphs and on the level of the transformation steps, enabling
a very powerful and flexible analysis method. This has been implemented
in the graph-based verification tool GROOVE. As an application of this
approach, we show how it gives rise to a competitive analysis technique
in the domain of feature modelling.

Keywords: Graph exploration analysis, Prolog, GROOVE, feature modelling.

1 Introduction

The practical value of graph transformation (GT) is especially determined by the
fact that graphs are a very general, widely applicable mathematical structure.
Virtually every artefact can be understood in terms of entities and relations
between them, which makes it a graph; and consequently, changes in such an
artefact can be specified through GT rules.

On the other hand, capability does not automatically imply suitability. For
instance, though it is possible to express structural properties as (nested) graph
conditions – see, for instance, [19,13] – in practice, if one wants to query a given
structure, writing graphical conditions to express and test for such queries is
not always the most obvious or effective way to go about it. This is particularly
true if the queries have not been predefined but are user-provided. Instead, there
are dedicated languages suitable for querying relational structures, such as, for
instance, SQL or Prolog.

---

* The work of this author is supported by the GRAIL project, funded by NWO (Grant
612.000.632).

A. Schürr, D. Varró, and G. Varró (Eds.): AGTIVE 2011, LNCS 7233, pp. 105–120, 2012.

The need for a powerful and flexible query language becomes even more clear when one wants to combine static (structural) properties with dynamic ones, so as to include the future or past evolution of the structure. For instance, temporal logic has been especially introduced to express dynamic properties and check them efficiently (see' [2] for an overview). However, besides lacking accessibility, temporal logic is *propositional*, meaning that it takes structural properties as basic building blocks; there is very little work on logics that can freely mix static and dynamic aspects of a system.

An example domain that requires this combination of static and dynamic aspects is *feature modelling*. A feature model is a graph in which nodes represent possible features (of some system under design) and edges express that one feature requires another, is in conflict, or is related in some other way. Graph transformation can be used to actually select features (in such a way that the constraints are met). The outcome is a (partially) resolved model, the quality of which is not only determined by the choices actually made but also by the possible choices still remaining. Thus, one would like to query a feature model for both its static properties (the choices actually made) and for its dynamic properties (the potential further transformation steps).

In this paper, we describe how one can use Prolog to query static and dynamic properties of graphs, simultaneously and uniformly. Besides the transformed graphs this requires a graph transition system (GTS), which is itself a graph with nodes corresponding to state graphs and edges to rule applications. The basic building block of Prolog is a *predicate*, which expresses a relation between its arguments. Example predicates in our setting are:

- The relation between a graph and its nodes or edges;
- The relation between an edge and its source or target node, or its label;
- The relation between a state of the GTS and its corresponding graph;
- The relation between one state of the GTS and the next.

A collection of Prolog predicates forms a *knowledge-base*, which is queried during the analysis of a GTS. Using an extension of the transformation tool GROOVE that supports Prolog queries, we demonstrate the capabilities of this approach on a case study based on feature modelling. This domain was chosen due to its applicability on the development process of software industries.

The paper is organised as follows. We first present the basic concepts for querying graphs using Prolog (Section 2); then we describe the application to feature modelling in Section 3. An analysis of the results can be found in Section 4. Conclusion and ideas for future work are given in Section 5.

2   Prolog in GROOVE

The Prolog programming language [7] is the *de facto* representative of the logic programming paradigm. Unlike imperative languages, Prolog is declarative: a

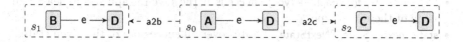

Fig. 1. Example GTS with three states and two transitions

Prolog program is composed of predicates about objects and their relations, and computations are performed by running queries over predicates. Given a query asking whether a predicate holds for a certain (given) object, the Prolog interpreter uses a *resolution* procedure that yields a yes or no answer. On the other hand, if a query has free variables, the Prolog engine will enumerate all objects which can be assigned to the variables so as to make the predicate true.

GROOVE [22,11] is a graph transformation tool set which can recursively explore and collect all possible rule applications over a start graph: this is referred to as the *exploration of the state space* of a graph grammar. The state space is stored as a graph transition system (GTS), where each state of the system contains a graph and each transition is labelled by a rule application. GROOVE has a graphical interface called the Simulator, for editing graphs and rules, and for exploring and visualising the GTS. The main technical contribution of this paper is the integration of a Prolog interpreter into the GROOVE Simulator.

2.1 Functionality Overview

Before executing Prolog queries, the standard GROOVE functionality is used to perform a state space exploration of the graph grammar under analysis. The exploration produces a GTS, which can then be inspected in queries. At this point it is important to stress the difference between *states* and *state graphs*. A state is an element of the GTS; it is implemented as an object with a unique identity and an associated state graph. A state graph is a host graph over which the transformation rules are applied.

We illustrate the Prolog functionality on the basis of a very small example. Figure 1 shows a GTS with three states, represented by dashed boxes: the start state s_0 and two successor states s_1 and s_2. Each of the states contains a state graph, consisting of two nodes connected by an e-labelled edge. The state graph of s_1 is obtained from the state graph of s_0 by applying rule a2b (not shown here) which renames an **A**-node to a **B**-node. Analogously, the state graph of s_2 is produced by applying rule a2c. Now consider the following Prolog query:

```
?− state(X), state_graph(X,GX), has_node_type(GX,'A'),
     state_next(X,Y), state_graph(Y,GY), has_node_type(GY,'C').
```

The query is composed of six predicates, interpreted conjunctively from left to right (the meaning of characters + and ? will be discussed in Section 2.2):

- state(?State) iterates over the states of the currently explored GTS.
- state_graph(+State, ?Graph) binds the state graph of the given state to the second argument; *i.e.*, it retrieves the state graph associated with the given state.

- has_node_type(+Graph, +Type) succeeds if the given graph has at least one node of the given type.
- state_next(+State, ?NextState) iterates over all successors of the given state.

The purpose of the query is to search for a state (variable Y) with a graph (GY) that has at least one node of type **C** and that has a predecessor state (X) whose graph (GX) contains a node of type **A**. Running this query produces the following result, which correctly binds Y to state s_2:

```
X = s0
GX = Nodes: [n0, n1]; Edges: [n0--A-->n0, n1--D-->n1, n0--e-->n1]
Y = s2
GY = Nodes: [n0, n1]; Edges: [n0--C-->n0, n1--D-->n1, n0--e-->n1]
Yes
More?
No
```

The output also shows the bindings for the other variables in the query. The values printed for variables GX and GY are the toString representations of the bound graphs, which show their internal structure – this explains the edge lists with three elements.[1] In the last two lines of the listing above, the user asked the interpreter if there are more results for the query. Since there are no other states that satisfy the query constraints, the answer is negative. If the GTS had more states satisfying the query, continuing the execution would eventually produce all of them. This is a consequence of the Prolog resolution procedure, which backtracks to predicate state_next, binding Y with other successors of X, as well as to state, binding X with other states of the GTS.

In addition to using the built-in GROOVE predicates, users can also define their own Prolog predicates. This ability to expand the Prolog knowledge-base (illustrated on Section 3.3) improves the extensibility of the framework.

2.2   Implementation Overview

Figure 2 shows the main elements of the integration of Prolog into the Simulator. GROOVE is written in Java, so in order to ease the coupling, we chose the GNU Prolog for Java library[2] [10] as our Prolog interpreter. The Simulator state in Figure 2 stands for the current snapshot of the Simulator configuration in memory. It contains Java objects that represent, among others, host graphs, transformation rules, and the GTS. The main block of Figure 2 is the glue code, which connects the Prolog interpreter to the rest of the Simulator. The glue code registers itself in the interpreter and is called back when a Prolog query is run. When called, the glue code inspects the Simulator state and tries to bind the Java objects with terms (variables) of the query.

---

[1] GROOVE uses an internal graph representation where nodes have very little structure; node types and flags are stored as special self-edges.

[2] http://www.gnu.org/software/gnuprologjava/

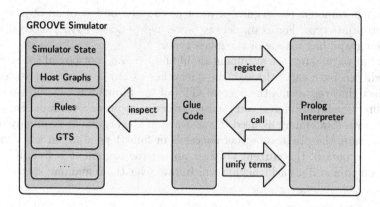

Fig. 2. Integration of the Prolog interpreter in the GROOVE Simulator

Built-in Predicates. Each built-in GROOVE predicate requires some glue code, written partly in Prolog and partly in Java. When the Prolog interpreter is created, an initialisation phase registers the built-in predicates with the interpreter. For instance, gts(−GTS) is a built-in predicate that binds the Java GTS object to a Prolog variable. Predicate registration is done with the following query:

```
:− build_in(gts/1, 'groove.prolog.builtin.Predicate_gts').
```

Predicate build_in is a special interpreter command for creating new predicates. The first argument specifies the predicate name and arity, the second one gives the name of the Java class that implements the predicate functionality. Here is a simplified listing for the Java Predicate_gts class.

```
1  public class Predicate_gts extends PrologCode {
2    public int execute(Interpreter interpreter, boolean backtracking, Term[] args) {
3      GTS gts = getSimulatorState().getGTS();
4      if (gts == null) {
5        return FAIL;
6      }
7      return interpreter.unify(args[0], gts);
8    }
9  }
```

When gts(X) is evaluated in a query, the interpreter calls execute of Predicate_gts. The third argument of the method is an array of Prolog terms that corresponds to the arguments of the predicate — in this case, X. The method first inspects the Simulator state to retrieve the GTS object (line 3). If the object is **null** the query fails, otherwise the object is bound to X (line 7).

Argument Modes. In the above, we have specified predicate signatures in which the parameter names were prefixed with special characters. These indicate the interaction of the Prolog interpreter with arguments at that position:

+ Input parameter: the argument must already be bound to an object of the appropriate type. For example, has_node_type(+Graph,+Type) succeeds if the given graph has a node of the given type.
− Output parameter: the argument should be *free*, *i.e.*, not bound to an object; it will receive a value through the query. For example, gts(−GTS) assigns the object that represents the current GTS of the Simulator.
? Bidirectional parameter: can be used either as input or as output. For example, state(?State) may be used in two ways. If the argument is already bound to a state, the predicate either succeeds or fails depending on whether that state is part of the current GTS or not. If the argument is free, it will be bound to a state; backtracking will iterate over the remaining states.

Backtracking. The Prolog resolution procedure is a search for valid bindings, in the course of which it may backtrack and re-evaluate predicates to retrieve further solutions. This implies that the implementation of the built-in predicates must handle backtracking. For example, the following is the Java glue code for predicate state_next.

```
1   public class Predicate_state_next extends PrologCode {
2     public int execute(Interpreter interpreter, boolean backtracking, Term[] args) {
3       PrologCollectionIterator it;
4       if (backtracking) {
5         it = interpreter.popBacktrackInfo();
6       } else {
7         State state = getSimulatorState().getState(args[0]);
8         it = new PrologCollectionIterator(state.getNextStateSet(), args[1]);
9         interpreter.pushBacktrackInfo(it);
10      }
11      return it.nextSolution();
12    }
13  }
```

The backtracking flag (line 2) is used by the interpreter to indicate if the predicate is being evaluated for the first time in a query or if it is being called again after backtracking. During the first run, the **else** block (lines 7–9) is executed. First the state object is retrieved along with its set of successor states (call to state.getNextStateSet()). This set is put into a special iterator along with the argument to be bound (line 8), which is then passed to the interpreter and stored as backtrack information (line 9). When the method is called again during backtracking, the same iterator is retrieved from the interpreter (line 5) and the next solution is returned.

3 Application to Feature Modelling

Feature models [15] are commonly used to support the configuration of products in software product lines [18]. They model variability by expressing commonalities, variations and constraints between the different features that could be part

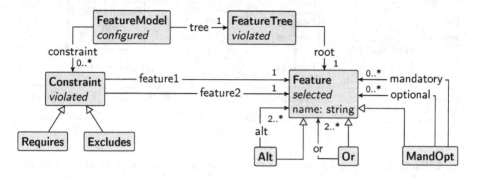

Fig. 3. A type graph for feature models

of a product. A feature usually represents an aspect of the software in an early phase of the software life cycle, and the impact of the combination of features is propagated across the phases until the actual product is implemented.

The analysis of feature models [5,25] is mostly concerned with verifying their static properties with respect to allowed specifications and valid configurations of the model. However, the specification of feature models and their configuration process go beyond the information in the model: they often involve multiple groups with distinct interests and expertise, which informally express extra properties of the features. Moreover, the definition of possible products depends on forces like market demands, user preferences, and the availability of assets at a specific time (such as the software components for the related products). Thus, feature modelling is a domain which can strongly profit from the ability to define and query static and dynamic properties of models, leading to richer analysis techniques. In particular, we can identify the following tasks in the analysis:

1. Model additional knowledge about features;
2. Define domain properties independently on the models, in a declarative way;
3. Simulate the configuration process;
4. Query for valid configurations with respect to conditions not expressed in the feature model;
5. Analyse alternative configuration paths and investigate the evolution of configuration stages.

We proceed to show how the Prolog extension for GROOVE can be used to implement these tasks. First we give an overview of the relevant concepts in terms of a type graph, some example rules and a small example model; then we focus on the use of Prolog to query the resulting state space of the grammar.

3.1   Feature Model Type Graph

Figure 3 shows a type graph for feature models (in GROOVE), based on the definitions given in [25]. The type **FeatureModel** represents a feature model composed of two parts: a **FeatureTree** whose nodes represent **Feature**s, and a set

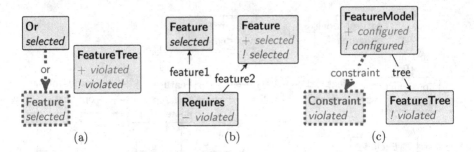

Fig. 4. Examples of graph production rules for feature model configuration. GROOVE rules are represented in a single graph, with different colours and line strokes used to distinguish different elements: black (continuous thin) elements are matched and preserved, and red (dashed fat) elements are Negative Application Conditions (NACs). Node flags preceded by a character have especial roles: + indicates flag creation; − is flag deletion, and ! is a NAC on the flag.

of explicit **Constraints** between these features. The constraint **Requires** indicates that if the target node of feature1 is selected for a product, then the target node of feature2 should be selected as well. The constraint **Excludes** indicates that the target nodes of feature1 and feature2 cannot be both selected for the same product. Type **Feature** has three subtypes: **MandOpt**, **Or** and **Alt**. The edges from each of these subtypes to a **Feature** indicate which kinds of child features each subtype can have. Leaf features of the tree are **MandOpt** features without children. Finally, the flags *configured*, *violated*, and *selected* in the type graph are used in GT rules to assist the configuration process and to enforce the identification of valid configurations.

3.2 Product Configuration

A specific feature model is a graph instantiating Figure 3, initially without any flags. The model is then configured using GT rules that encode the following constraints (some of which were discussed above):

1. The root feature must be selected first;
2. When a **MandOpt** is selected, all mandatory children must also be selected;
3. When an **Or** is selected, at least one of its children must also be selected;
4. When an **Alt** is selected, exactly one of its children must also be selected;
5. When a non-root feature is selected, its parent feature must also be selected.

Child features are selected on demand and violations of constraints are checked at each step. This applies both to the implicit conditions of the **FeatureTree** and to the explicit **Requires** and **Excludes** constraints in the model.

Each of the steps above is performed by a combination of graph transformation rules. For example, Figure 4(a) shows a rule used to detect a violation on the selection of a child feature of an **Or** (step 3); the rule in Figure 4(b) selects a feature which is required by another, previously selected one and removes the violation of the **Requires** constraint; and the rule in Figure 4(c) checks the conditions for the complete feature model to be correctly configured and marks it as *configured*. A valid configuration of the feature model is found when neither the constraints nor the feature tree are violated. Note that a tree violation is modelled independently of the violation of explicit constraints between pair of features. A tree has a violation when one of the requirements listed above is not satisfied, *e.g.*, when the root feature is not selected. Each valid configuration selects a set of features that gives rise to a potential product of the product line.

Starting from the initial feature model, state space exploration generates a GTS resulting from all possible interleavings of rule applications. Each state represents the feature model with a partial selection of features, some of which may form valid configurations. Figure 5 shows a completely configured feature model, immediately after the application of the rule shown in Figure 4(c) (named FeatureModelConfiguration in the grammar). This configuration has a set of *selected* features and no constraint violations (*violated* flags). Note that the mandatory part of feature EnergySaving does not have to be selected since the feature itself, which is optional, was not selected. Once generated, the GTS can be queried using Prolog.

3.3   Querying the State Space

We now come to the main point of the example, which is how Prolog may be used to analyse the state space. For instance, the following user-defined predicate extracts completely configured products:

```
product(Product) :−
 rule('FeatureModelConfiguration', Rule), % Get the rule object
 % Get the graph resulting from rule application
 rule_application_result(Rule, Graph),
 % Collect all features selected to compose the product.
 findall(Feature, selected_feature(Graph, Feature), Product).
```

The predicate searches for graphs resulting from the application of rule FeatureModelConfiguration and then collects all selected features in this graph (using **findall**, which is a higher-order predicate provided by GNU Prolog). Successive calls of product generate all valid models. For the initial, unconfigured version of the feature model this yields 50 products, including the one shown in Figure 5 (composed of features HomeAuto, Surveil, AccidentDet, AlarmAuto, Alarm, and Bell). Predicate product uses the following auxiliary predicate, which consults information of the GTS.

```
rule_application_result(Rule, Graph) :−
 state(Source), % Get a source state
 state_transition(Source, Transition), % Get a transition from source state
 transition_event(Transition, Event), % Get the rule application event of the transition
 ruleevent_rule(Event, Rule), % Ensure that the given rule is the one that was applied
 transition_target(Transition, Target), % Get the target state of the transition
 state_graph(Target, Graph). % Get the graph of target state
```

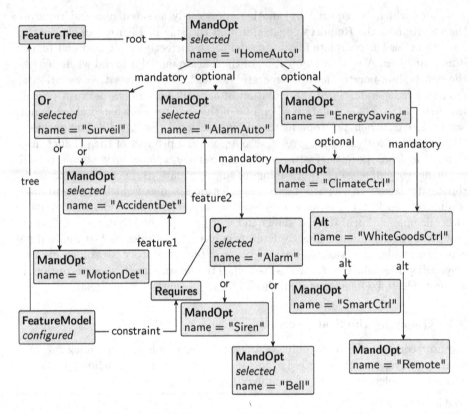

Fig. 5. A valid configuration of the feature model

This predicate uses the rule application event associated with transitions of the GTS to ensure that the given rule is the one that was indeed applied in the Transition. It is important to note that the GTS has a total of 312 states, representing all intermediate states that lead to one of the 50 configured feature models. These intermediate states allow the analysis of different evolution paths (formed by different rule application sequences) that lead to the same solution. Furthermore, the grammar contains rules to re-validate constraint violations, making it possible to reach a valid configuration even if a constraint was violated in an intermediate state.

Another useful capability provided by the Prolog extension is the possibility to define a knowledge base of additional model-related information. As an example, suppose that we are interested in products that satisfy a certain budget constraint. The following Prolog code sets the costs for each feature of the model and defines what it means for a product to be within budget.

```
% Extra facts about feature costs.
cost('HomeAuto', 1). cost('Surveil', 0). cost('WhiteGoodsCtrl', 10).
cost('AlarmAuto', 10). cost('EnergySaving', 5). cost('Siren', 15).
```

```
cost('AccidentDet', 25). cost('ClimateCtrl', 10). cost('Bell', 10).
cost('MotionDet', 25). cost('Alarm', 5). cost('SmartCtrl', 10). cost('Remote',10 ).

% Computes the cost of a product.
sum_costs([], 0).
sum_costs([H|T], Total) :− sum_costs(T, CT), cost(H, CH), Total is CH+CT.

% Checks if the given product is within the given budget.
within_budget(Budget, Product) :− sum_costs(Product, Cost), Cost =< Budget.
```

The following query returns products with total cost smaller or equal to 70:

```
?− product(P), within_budget(70, P).
```

For our running example, this gives 11 products within the budget constraint.

4 Discussion and Related Work

In the previous section we showed how the Prolog extension for GROOVE supports
the graph-based representation of feature models, how extra model attributes can
be specified as Prolog predicates and how state space exploration can be used
to search for feature model configurations. Going back to the list of tasks in
Section 3 (page 111), we see that, in fact, all of them are fulfilled.

4.1 Performance

There are two major points of evaluation regarding run-time performance of
the tool set: (i) the time used to explore the state space and store the GTS;
and (ii) the time needed to run the Prolog queries on a given GTS. Item (i)
covers the standard functionality of GROOVE, for which an extensive body of
work exists [21,20,8,12], comprising performance evaluations of several aspects
of GROOVE implementation and also comparisons with other tools. Item (ii)
concerns the functionality introduced by the Prolog extension, which requires a
new analysis.

For the running example of Section 3 the time necessary for building the GTS
and performing the Prolog queries is negligible (around 2 milliseconds in total),
since the state space is small. To properly exercise the tool, we used a grammar
implementing a solution for the leader election case study [16] proposed at the
GraBaTs 2009 tool contest. The purpose of this case is to verify a protocol for the
election of a leader among a ring of processes. The state space size is exponential
on the number of processes, giving rise to very large transition systems.

The results of the experiments are given in Table 1. Column # **States** shows
the GTS size for an increasing number of processes. Columns **Exploration
Time** and **Query Time** give the time in milliseconds needed to explore the
state space and run the Prolog queries, respectively. We used a query similar to
the one given in Section 3.3, that collects all states of the GTS where a leader has
been elected. The last column lists the number of results returned by the query.

Table 1. Performance comparison of Prolog queries against state space size.

# States	Growth Ratio	Exploration Time (ms)	Growth Ratio	Query Time (ms)	Growth Ratio	# Results
10		251		< 1		2
52	5.2	347	1.4	3	4.1	10
473	9.1	1,000	2.9	30	9.0	84
6,358	13.4	6,001	6.0	294	9.9	1,008
113,102	17.8	140,961	23.5	12,238	41.6	15,840

From the times given in Table 1 it can be seen that the bulk of the running time is spent on building the GTS, unsurprisingly. From the growth ratios we see that the query time increases linearly over the GTS size until the second-to-last line. However, at the last line of the table, the query time exhibits a larger growth, which can be explained by the large amount of backtracking done by the Prolog interpreter while running the query.

Although further experimentation is certainly in order, we consider these initial performance results of the Prolog extension satisfactory.

4.2   Related Feature Modelling Approaches

The analysis of feature models is useful for several reasons, such as to efficiently resolve the configuration constraints and to optimise the configuration calculation. This analysis is the object of research in many directions, which differ in the expressiveness of the models and in the configuration strategies. For example, generalised feature trees [25], propositional formulas [4] and constraint satisfaction problems [5] have been used for the purpose of analysing feature models. Although some of these algorithms are known to be quite efficient, the major drawback of such approaches is the rigidity of the analysis method. A review of the current techniques for the automated analysis of feature models is given in [6].

Extra feature attributes can be modelled in our approach in at least two ways: first, by adding new attributes to the graphs; or secondly, by defining predicates in Prolog that represent such attributes. We chose the last form because the values of the attributes used can be quite volatile in this application domain. Again, it is possible to annotate the graph with all kinds of information, but this would hamper the flexibility of the approach.

We want to draw attention to the issue of *staged configuration* of product lines. A stage corresponds to the elimination of a set of configuration options; the selection of features is deferred through stages until no variability is left. Czarnecki *et al.* [9] handle staged configuration using a feature model notation that supports the definition of feature cardinality. They explore the variability in a feature model per stage, which contains the features that can be selected. Hubaux *et al.* [14] propose a way to guide the configuration process using workflows which enforce the staged configuration in a certain order. Both approaches also handle inter-related feature models, in which the configuration order matters but is predefined and fixed over the whole configuration process.

We are able to generate all configuration stages of a feature model, as graph states, and to inspect these stages in several ways: by querying in which order the features (especially the variable features) are selected, or also by making several kinds of inspections in these stages. For example, our GROOVE solution supports the analysis of configuration contexts in which a constraint has been violated. We can also add extra constraints which are combinations of conditions in previous stages.

4.3   Related Tools

PROGRES [23] is a specification language which provides several mechanisms for defining graph properties, including derived attributes, restrictions and paths (unary and binary relations on nodes, which may be defined both textually and graphically), graphical queries (called graph tests in PROGRES), and constraints (structural conditions going beyond the expressive facilities of graph schemas). In contrast to the GROOVE/Prolog integration, PROGRES does not support ad-hoc queries on graphs, and it also does not support queries on graph transition systems.

Among other tools for the verification of graph transformation (GT) systems we can cite AUGUR2 [17] and ENFORCe [1]. AUGUR2 uses abstraction to verify GT grammars with infinite state spaces. ENFORCe acts as proof checker for the correctness high-level programs written as graph transformations. While both these tools could be used to analyse the evolution of a graph to some degree, the explicit-state model checking approach of GROOVE gives an advantage, since it provides a simpler representation of intermediate states that eases the understandability for the layman user. For a more comprehensive comparison between GROOVE and other GT tools see [11].

Concerning other existing combinations of graph transformation tools with Prolog, as far as we are aware, there are only two similar approaches, embodied by VIATRA2 [26] and VMTS [27], both of which are GT-based tools for model transformation.

VIATRA2. Varró and Balogh [3] describe how VIATRA2 and Prolog can be used to implement their so called Model Transformation by Example (MTBE) approach.

The purpose of MTBE is to semi-automatically derive model transformation rules from example relations between source and target model elements. These example relations are represented in VIATRA2 using a mapping model, formed by the source and target meta-models and a reference meta-model to interconnect them. The mapping model is translated to Prolog clauses and an inductive learning program is run, producing Prolog inference rules representing hypothesis that are satisfied under the given clauses. These inference rules are then translated back to a VIATRA2 representation and give rise to model transformation rules that can operate on instances of the source and target meta-models, following the example relations given in the mapping model. This process can be repeated in order to iteratively refine the rules produced.

From the above, it should be clear that the intended use of Prolog in the setting of VIATRA2 is quite different from ours, and hence there is little basis for a deeper comparison.

VMTS. At the GraBaTs 2009 tool contest, Siroki *et al.* [24] presented a solution to the leader election case study using VMTS and Prolog.

The goal of their approach is to check if the outcome of a set of model transformation rules applied to a given input model complies to certain properties. To perform this analysis, first the input model, the transformation rules and the control flow graph specifying the order for rule applications are all translated from the VMTS format to a Prolog representation. Subsequently, the Prolog resolution procedure is used to enumerate the possible output models of the transformation. Finally, these output models are checked by Prolog predicates that express the properties one wants to assert.

Their use of Prolog resolution plays the same role as the state space exploration functionality of GROOVE. However, their approach suffers from the need to translate VMTS objects to Prolog. The Prolog resolution procedure is not adequate for the exploration of a graph-based state space and therefore gives poor performance. Another consequence of the translation is the low readability of the generated Prolog clauses.

5   Conclusions and Future Work

Summarising, the highlights of the approach described in this paper are:

- Prolog is tightly integrated with graph-based state space exploration;
- Queries can uniformly combine static and dynamic aspects of graphs;
- The framework supports user-defined Prolog facts and predicates.

We have demonstrated these advantages by applying the approach in the domain of feature modelling, where it gives rise to a competitive alternative to existing, more rigid frameworks.

We have implemented the above as an extension to GROOVE. Although many of the examples given in this paper could have been solved in GROOVE using other means, the Prolog-based solutions are more convenient and elegant. Therefore, the extension improves usability, which is a key factor for success.

On a more general level, this paper shows that there is much to be gained when graph transformation is connected to other techniques, and that this connection can be done in a simple, uniform way.

Future Work. There are two main points planned as future work.

- Prolog-*based application conditions.* One can associate Prolog queries to individual GT rules, to play the role of additional application conditions. When a rule with a query is matched, the query is executed in the Prolog interpreter, and only if the query succeeds the rule is applied. This functionality is orthogonal to other application conditions already present in GROOVE, such as NACs, and would give another option for controlling the flow of rule applications, in addition to rule priorities and control programs.

- Prolog-*based state space exploration*. One can also extend the GROOVE exploration strategies with a condition based on a Prolog query. Every time a new state is produced, the query is run, and if the query is successful the state is added to the GTS. The effect is comparable to a global post-application condition.

Availability. The Prolog extension described in this paper is implemented in GROOVE version 4.4.0, available at http://groove.cs.utwente.nl. The grammar for the solution given in Section 3 can also be downloaded at the same address.

Acknowledgement. The integration of Prolog into GROOVE is originally due to Michiel Hendriks.

References

1. Azab, K., Habel, A., Pennemann, K.H., Zuckschwerdt, C.: ENFORCe: A system for ensuring formal correctness of high-level programs. In: Zündorf, A., Varró, D. (eds.) Proc. of the 3rd Int. Workshop on Graph-Based Tools. ECEASST, vol. 1 (2007)
2. Baier, C., Katoen, J.P.: Principles of Model Checking. MIT Press (2008)
3. Balogh, Z., Varró, D.: Model transformation by example using inductive logic programming. Software and System Modeling 8(3), 347–364 (2009)
4. Batory, D.: Feature Models, Grammars, and Propositional Formulas. In: Obbink, H., Pohl, K. (eds.) SPLC 2005. LNCS, vol. 3714, pp. 7–20. Springer, Heidelberg (2005)
5. Benavides, D., Trinidad, P., Ruiz-Cortés, A.: Automated Reasoning on Feature Models. In: Pastor, Ó., Falcão e Cunha, J. (eds.) CAiSE 2005. LNCS, vol. 3520, pp. 491–503. Springer, Heidelberg (2005)
6. Benavides, D., Segura, S., Ruiz-Cortés, A.: Automated analysis of feature models 20 years later: A literature review. Information Systems 35, 615–636 (2010)
7. Clocksin, W.F., Mellish, C.S.: Programming in Prolog. Springer (1984)
8. Crouzen, P., van de Pol, J.C., Rensink, A.: Applying formal methods to gossiping networks with mCRL and GROOVE. In: Haverkort, B.R.H.M., Siegle, M., van Steen, M. (eds.) ACM SIGMETRICS Performance Evaluation Review, vol. 36, pp. 7–16. ACM, New York (2008)
9. Czarnecki, K., Helsen, S., Eisenecker, U.: Staged Configuration Using Feature Models. In: Nord, R.L. (ed.) SPLC 2004. LNCS, vol. 3154, pp. 266–283. Springer, Heidelberg (2004)
10. Diaz, D., Codognet, P.: The GNU Prolog system and its implementation. In: ACM Symposium on Applied Computing (SAC), vol. 2, pp. 728–732. ACM, New York (2000)
11. Ghamarian, A., de Mol, M., Rensink, A., Zambon, E., Zimakova, M.: Modelling and analysis using GROOVE. International Journal on Software Tools for Technology Transfer (STTT) (March 2011)

12. Ghamarian, A.H., Jalali, A., Rensink, A.: Incremental pattern matching in graph-based state space exploration. In: de Lara, J., Varró, D. (eds.) Proc. of the 4th Int. Workshop on Graph-Based Tools. ECEASST, vol. 32 (2010)
13. Habel, A., Pennemann, K.-H., Rensink, A.: Weakest Preconditions for High-Level Programs. In: Corradini, A., Ehrig, H., Montanari, U., Ribeiro, L., Rozenberg, G. (eds.) ICGT 2006. LNCS, vol. 4178, pp. 445–460. Springer, Heidelberg (2006)
14. Hubaux, A., Classen, A., Heymans, P.: Formal modelling of feature configuration workflows. In: Muthig, D., McGregor, J.D. (eds.) Software Product Lines Conference (SPLC). ACM International Conference Proceeding Series, vol. 446, pp. 221–230. ACM (2009)
15. Kang, K.C., Cohen, S.G., Hess, J.A., Novak, W.E., Peterson, A.S.: Feature-oriented domain analysis (FODA) feasibility study. Tech. rep., Carnegie-Mellon University Software Engineering Institute (November 1990)
16. König, B.: Case Study: Leader Election, http://is.tm.tue.nl/staff/pvgorp/events/grabats2009/cases/grabats2009verification.pdf
17. König, B., Kozioura, V.: Augur 2 – A new version of a tool for the analysis of graph transformation systems. In: Bruni, R., Varró, D. (eds.) Proc. of the 5th International Workshop on Graph Transformation and Visual Modeling Techniques. ENTCS, vol. 211, pp. 201–210. Elsevier (2008)
18. Pohl, K., Böckle, G., van der Linden, F.J.: Software Product Line Engineering: Foundations, Principles and Techniques. Springer-Verlag New York, Inc., Secaucus (2005)
19. Rensink, A.: Representing First-Order Logic Using Graphs. In: Ehrig, H., Engels, G., Parisi-Presicce, F., Rozenberg, G. (eds.) ICGT 2004. LNCS, vol. 3256, pp. 319–335. Springer, Heidelberg (2004)
20. Rensink, A.: Isomorphism checking in GROOVE. In: Zündorf, A., Varró, D. (eds.) Proc. of the 3rd Int. Workshop on Graph-Based Tools. ECEASST, vol. 1 (2007)
21. Rensink, A., Schmidt, Á., Varró, D.: Model Checking Graph Transformations: A Comparison of Two Approaches. In: Ehrig, H., Engels, G., Parisi-Presicce, F., Rozenberg, G. (eds.) ICGT 2004. LNCS, vol. 3256, pp. 226–241. Springer, Heidelberg (2004)
22. Rensink, A.: The GROOVE Simulator: A Tool for State Space Generation. In: Pfaltz, J.L., Nagl, M., Böhlen, B. (eds.) AGTIVE 2003. LNCS, vol. 3062, pp. 479–485. Springer, Heidelberg (2004)
23. Schürr, A., Winter, A.J., Zündorf, A.: The PROGRES approach: Language and environment. In: Ehrig, H., Engels, G., Kreowski, H.J., Rozenberg, G. (eds.) Handbook of Graph Grammars and Computing by Graph Transformation, pp. 487–550. World Scientific Publishing Co., Inc., River Edge (1999)
24. Siroki, L., Vajk, T., Madari, I., Mezei, G.: VMTS Solution of Case Study: Leader Election, http://is.tm.tue.nl/staff/pvgorp/events/grabats2009/submissions/grabats2009_submission_18-final.pdf
25. van den Broek, P.M., Galvao, I.: Analysis of feature models using generalised feature trees. In: Workshop on Variability Modelling of Software-Intensive Systems, No. 29 in ICB-Research Report, Universität Duisburg–Essen, Germany, pp. 29–35 (January 2009)
26. VIATRA2– Visual Automated Model Transformations Framework, http://www.eclipse.org/gmt/VIATRA2/
27. VMTS– Visual Modeling and Transformation System, http://vmts.aut.bme.hu/

Graph Grammar Induction
as a Parser-Controlled Heuristic Search Process

Luka Fürst[1], Marjan Mernik[2], and Viljan Mahnič[1]

[1] University of Ljubljana, Faculty of Computer and Information Science,
Tržaška cesta 25, SI-1000 Ljubljana, Slovenia
{luka.fuerst,viljan.mahnic}@fri.uni-lj.si
[2] University of Maribor, Faculty of Electrical Engineering and Computer Science,
Smetanova ulica 17, SI-2000 Maribor, Slovenia
marjan.mernik@uni-mb.si

Abstract. A graph grammar is a generative description of a graph language (a possibly infinite set of graphs). In this paper, we present a novel algorithm for inducing a graph grammar from a given set of 'positive' and 'negative' graphs. The algorithm is guaranteed to produce a grammar that can generate all of the positive and none of the negative input graphs. Driven by a heuristic specific-to-general search process, the algorithm tries to find a small grammar that generalizes beyond the positive input set. During the search, the algorithm employs a graph grammar parser to eliminate the candidate grammars that can generate at least one negative input graph. We validate our method by inducing grammars for chemical structural formulas and flowcharts and thereby show its potential applicability to chemical engineering and visual programming.

Keywords: Graph grammars, graph grammar induction, graph grammar parsing, heuristic search.

1 Introduction

Despite a large variety of applications [1,5,17], graph grammars have seldom been used for classifying, compressing, or characterizing graph sets. However, these potential roles would become far more important if grammars could be automatically induced from graphs. For example, by inducing a graph grammar from a set of chemical structural formulas, one could acquire a classifier to distinguish, e.g., biologically active substances from others, a way to compress large chemical databases, or a set of rules characterizing a given group of chemicals.

In this paper, we present a novel approach to inducing graph grammars from positive and (optionally) negative graph examples. Our algorithm is guaranteed to produce a grammar that can generate all of the positive and none of the negative examples. By formulating grammar induction as a best-first search process biased towards small grammars, the algorithm may be expected to induce a grammar that generalizes beyond the observed examples. The search proceeds in the specific-to-general direction, starting with a trivial grammar that can

A. Schürr, D. Varró, and G. Varró (Eds.): AGTIVE 2011, LNCS 7233, pp. 121–136, 2012.

generate exactly the positive input graph set. As the algorithm progresses, it produces increasingly smaller and more general candidate grammars. To prevent over-generalization, the algorithm is coupled with a graph grammar parser, which is used to check whether a given candidate grammar can generate any negative input graph. If it can, it is immediately discarded.

In the domain of graph grammar induction, only few approaches have been formulated as a parser-controlled search process. However, the main contribution of this paper is the generalization operator in the search process, i.e., the way of proceeding from more specific to more general grammars.

The grammars induced by our algorithm constitute a subclass of the Layered Graph Grammars (LGG) formalism [18] and can therefore be parsed using the Rekers-Schürr parser [18,8]. The parsability of the target formalism makes it possible to induce a grammar from both positive and negative examples, which results in a grammar suitable for classification purposes.

In this paper, we present two nontrivial applications of our algorithm. First, we induce a grammar of flowcharts comprising atomic, sequential, conditional, and iterative statements. As a second application, we induce a grammar of the structural formulas of a subset of hydrocarbons (chemical compounds comprising carbon and hydrogen atoms). The potential applications of inducing grammars from chemical formulas have already been mentioned. Induction of flowchart grammars (and diagram grammars in general) may find its uses in visual programming tools. Graph grammars are often difficult to create 'by hand'. Using our approach, a tool could automatically induce a parsable graph grammar from a few user-provided sample graphs.

The rest of this paper is structured as follows: In Sect. 2, we give a review of related work. Section 3 defines the basic concepts. Our approach is described in Sect. 4 and experimentally validated in Sect. 5. Section 6 concludes the paper.

2   Related Work

The work on graph grammar induction has been fairly scarce. This fact can be attributed partly to the complexity of the problem itself and partly to the lack of efficient general parsers, which stems from the NP-hardness of the parsing problem for many classes of graph grammars.

One of the first graph grammar induction approaches was proposed by Jeltsch and Kreowski [10]. Their algorithm induces a hyperedge replacement (HR) grammar [19, Chap. 2] from a set of positive graphs by successive generalizations of the trivial initial grammar. Our approach is based on a similar idea, but we employ a fairly different generalization operator and embed the generalization scheme into a search algorithm.

Jonyer et al. [11] also induce grammars from positive graphs in a specific-to-general direction. In each generalization step, their algorithm determines the 'best' (according to the Minimum Description Length principle) subgraph S in the input set, replaces it with a single nonterminal vertex v, and adds the production $v ::= S$ to the grammar. The generated productions are not equipped with any embedding rules and can therefore only represent chains of similar graphs connected with single edges. An improved version of this approach, proposed by Kukluk et al. [13], induces grammars that can represent sequences of graphs sharing common edges. Recently, Brijder and Blockeel [4] presented a method to induce a node-label controlled (NLC) grammar [19, Chap. 1] from a single graph containing a set of isomorphic subgraphs.

None of the approaches mentioned above makes use of a parser. Therefore, they cannot accept negative graphs, and the induced grammars are not suitable for classification purposes. An approach that does employ a parser, although only to validate the final grammar produced by the induction algorithm, was proposed by Ates et al. [2]. They induce grammars from the Spatial Graph Grammar formalism [12], which is parsable in polynomial time but fairly restricted.

Our approach induces grammars that are both parsable and fairly powerful, at least in comparison to those of Jonyer et al. and Ates et al. Another advantage of our method is that the parser actively participates in the induction process. Unfortunately, the combination of the power and parsability of the target formalism results in the exponential worst-case complexity of the parser and hence of the entire algorithm.

The problem of graph grammar induction has been inspired by that of string grammar induction [16], where many approaches are based on similar ideas as our method, i.e., specific-to-general search, parser-based validation, etc. [7,15].

Graph grammar induction is also related to the problem of metamodel inference [9], where the goal is to induce a metamodel from a given set of models, and to that of model transformation by example [3], where the goal is to infer model transformation rules from a set of known transformation pairs. Since model transformation rules can be represented as graph grammars in the Triple Graph Grammar (TGG) formalism [20], the model-transformation-by-example problem can be formulated as a TGG induction problem.

3   Definitions

A *directed graph* G is a tuple $(V_G, E_G, VLabels_G, ELabels_G, conn_G, vlabel_G, elabel_G)$, where V_G, E_G, $VLabels_G$, and $ELabels_G$ are the sets of *vertices*, *edges*, *vertex labels*, and *edge labels*, respectively, $conn_G : E_G \rightarrow V_G \times V_G$ is the function defining the source and the target vertex for each edge, and $vlabel_G : V_G \rightarrow VLabels_G$ and $elabel_G : E_G \rightarrow ELabels_G$ are the functions defining the *labels* of individual graph elements. For convenience, let $label_G(x) \equiv vlabel_G(x)$ if $x \in V_G$ and $label_G(x) \equiv elabel_G(x)$ if $x \in E_G$. Unlabeled vertices and edges will be treated as if they were labeled with a special label ϕ_V and ϕ_E, respectively. Let $|G| = |V_G| + |E_G|$ be the *size* of the graph G. *Undirected graphs* are defined in the same way as directed ones, except that for each edge e, $conn(e)$ is a two-element set rather than an ordered pair. The subscripts in V_G, E_G, $conn_G$, etc., will be omitted when the associated graph is clear from context.

Graphs G and H are *isomorphic* (denoted $G \approx H$) if there exists a bijective vertex-to-vertex and edge-to-edge mapping (called *isomorphism*) $h : G \rightarrow$

H that preserves labels and adjacencies, i.e., $label_H(h(x)) = label_G(x)$ and $conn_H(h(e)) = h(conn_G(e))$ for all $x \in V_G \cup E_G$ and $e \in E_G$.[1] A graph H is a *subgraph* of a graph G (denoted $H \preceq G$) if $V_H \subseteq V_G$ and $E_H \subseteq E_G$. An *occurrence* of a graph H in a graph G is a subgraph $H' \preceq G$ such that $H' \approx H$. Let us define the *neighborhood* of a subgraph $H \preceq G$ in G (denoted $Nh_G(H)$) as the set of all vertices in $V_G \setminus V_H$ connected to at least one vertex in V_H, i.e., $Nh_G(H) = \{v \in V_G \setminus V_H \mid \exists w \in V_H, e \in E_G : conn_G(e) = (v, w) \vee conn_G(e) = (w, v) \vee conn_G(e) = \{v, w\}\}$.

Let $[u: A \xrightarrow{e:\,t} v: B]$ (or $[u: A \overset{e:\,t}{\text{---}} v: B]$) denote a graph comprising a vertex u labeled A, a vertex v labeled B, and an edge e labeled t with $conn(e) = (u, v)$ (or $conn(e) = \{u, v\}$). Let $[u(S)v]$ denote a graph comprising vertices u and v, a subgraph S such that $Nh_{[u(S)v]}(S) = \{u, v\}$, and an arbitrary number of edges connecting the vertices of S to the vertices u and v.

Let us now define the graph grammar formalism induced by our method. A *graph grammar* is the quadruple $GG = (\mathcal{T}^V, \mathcal{T}^E, \mathcal{N}^E, \mathcal{P})$, where $\mathcal{T}^V$, $\mathcal{T}^E$, and $\mathcal{N}^E = \{\#1, \#2, \ldots\}$ are pairwise disjoint sets of *terminal vertex labels*, *terminal edge labels*, and *nonterminal edge labels*, respectively, and $\mathcal{P}$ is a set of *productions* of the form $p: L ::= R$, where $L = LHS(p)$ (the left-hand side or LHS) and $R = RHS(p)$ (the right-hand side or RHS) are connected graphs such that $VLabels_L \subseteq VLabels_R \subseteq \mathcal{T}^V$, $ELabels_L \subseteq \mathcal{N}^E$, and $ELabels_R \subseteq \mathcal{T}^E \cup \mathcal{N}^E$. Additionally, each production has to belong to one of the following types:

Type I: Productions of this type take the form $\lambda ::= R$, where λ denotes the graph with no elements (the null graph).

Type II: These productions take the form $[u: A \xrightarrow{e:\,\#i} v: B] ::= [u(S)v]$ or $[u: A \overset{e:\,\#i}{\text{---}} v: B] ::= [u(S)v]$, where $\{A, B\} \subseteq \mathcal{T}^V$ and $\#i \in \mathcal{N}^E$. The subgraph S will be called the *core*, and the vertices u and v will be called the *guards*. Productions of this type will often be written as $[A \xrightarrow{\#i} B] ::= [A(S)B]$ or $[A \overset{\#i}{\text{---}} B] ::= [A(S)B]$.

Type III: These productions take the form $[u: A \xrightarrow{e:\,\#i} v: B] ::= [u \xrightarrow{r} v]$ or $[u: A \overset{e:\,\#i}{\text{---}} v: B] ::= [u \overset{r}{\text{---}} v]$, where $\{A, B\} \subseteq \mathcal{T}^V$, $\#i \in \mathcal{N}^E$, and $r \in \mathcal{T}^E \cup \mathcal{N}^E$. Productions of this type will often be written as $[A \xrightarrow{\#i} B] ::= [A \xrightarrow{r} B]$ or $[A \overset{\#i}{\text{---}} B] ::= [A \overset{r}{\text{---}} B]$.

For example, the grammar GG_7 in Fig. 2 contains one production of each type ($p_{7,1}$ belongs to type I, $p_{7,2}$ to type II, and $p_{7,3}$ to type III). The guard vertices on production RHSs are marked with small black circles. In the case of directed grammars (e.g., in Fig. 1), the RHS guards are marked with 'S' and 'T'. The letter 'S' marks the vertex that coincides with the source vertex on the LHS.

To *apply* a type-II production $p: [u: A \xrightarrow{e:\,\#i} v: B] ::= [u(S)v]$ to a graph G, one has to (1) find an occurrence L' of the graph $[u: A \xrightarrow{e:\,\#i} v: B]$ in G, (2) replace in L' the edge that corresponds to e with a copy S' of the graph S, and

---

[1] For any function $f: A \to B$, let $f((x, y)) = (f(x), f(y))$ and $f(\{x, y\}) = \{f(x), f(y)\}$.

(3) connect the vertices of S' to the vertices corresponding to u and v in the same way as the vertices of S are connected to the vertices u and v in $RHS(p)$. To apply a type-III production $[u: A \xrightarrow{e:\ \#i} v: B] ::= [u \xrightarrow{r} v]$ to a subgraph $[u': A \xrightarrow{e':\ \#i} v': B]$, the edge e' has to be replaced with an edge labeled r. To *reverse-apply* a production, this procedure is simply reversed. Undirected productions are applied and reverse-applied in the same way as directed ones. A sample grammar and a series of production applications is shown in Fig. 1. The notation $L ::= R_1 \mid \ldots \mid R_s$ is an abbreviation for $L ::= R_1, \ldots, L ::= R_s$.

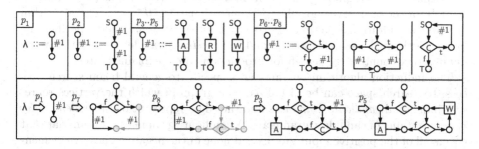

Fig. 1. *Top row*: The reference grammar for the flowcharts language. *Bottom row*: A derivation of a sample flowchart. The application of the production p_8 is highlighted.

A grammar GG *covers* a graph G if GG can generate G, i.e., if G can be derived from the null graph using the productions of GG. The *language* of a grammar GG is the set of all terminal-labeled graphs covered by GG. A *parser* is an algorithm that determines whether a given graph G belongs to the language of a given grammar GG.

The RHSs of the type-I productions of a grammar GG will be collectively called the *base graphs* of GG and denoted $\mathcal{B}(GG)$. The *size* of a grammar will be defined as $|GG| = \sum_{p \in \mathcal{P}(GG)} |p|$, where $|p| = |RHS(p)|$ if p is a type-I production, and $|p| = |RHS(p)| + |LHS(p)| - 2 = |RHS(p)| + 1$ otherwise. (The two guards are common to the LHS and RHS, hence '−2'.) A grammar GG_1 is *larger* (or *smaller*) than a grammar GG_2 if $|GG_1| > |GG_2|$ (or $|GG_1| < |GG_2|$).

Our target grammar formalism can be regarded as a subset of both LGG and HR formalisms. Hyperedge replacement grammars consist of productions for replacing individual hyperedges with hypergraphs. A *hyperedge* is an edge that connects an arbitrary sequence or multiset of vertices. (An ordinary edge is therefore a special case of a hyperedge.) A *hypergraph* is a graph composed of vertices and hyperedges. Type-I productions can be thus viewed as HR productions with zero-arity hyperedges on their LHSs. Type-II and type-III productions also specify HR rules, since the guard vertices do not participate in the replacement process itself; rather, they only determine the context of replacement. The guards correspond to *external vertices* in the HR terminology.

4   The Proposed Graph Grammar Induction Algorithm

In this section, we will often refer to Fig. 2, which shows the induction of a grammar from a single positive graph, namely the structural formula of butane. In this example, the final result is the grammar GG_7.

4.1   Overview

The pseudocode of the induction algorithm is shown in Fig. 3. The induction algorithm induces a graph grammar from a set of positive graphs ($\mathcal{G}^+$) and (optionally) a set of negative graphs ($\mathcal{G}^-$) such that $\mathcal{G}^+ \cap \mathcal{G}^- = \emptyset$. The algorithm accepts two additional parameters (positive integers): *beamWidth* specifies the beam width in the search process, and *maxVertexCount* determines the maximum vertex count in the search for production cores (explained later).

The induction algorithm operates as a specific-to-general beam search process. Its search space can be visualized as a graph in which the vertices represent individual *candidate grammars* and the edges represent possible *elementary generalizations* of candidate grammars. A candidate grammar is a grammar that covers all of the positive input graphs and none of the negative ones. An elementary generalization step transforms a given candidate grammar GG into a new candidate grammar that is at least as general as GG.

The goal of the algorithm is to find a candidate grammar of the minimum size. By restricting its search to the space of candidate grammars, the algorithm is guaranteed to produce a correct grammar in terms of the coverage of the input graphs. Its preference for small grammars is likely to result in a grammar that generalizes beyond the observed examples.

The algorithm starts with the most specific candidate grammar. This grammar, denoted GG_1, consists of productions $\{\lambda ::= G \mid G \in \mathcal{G}^+\}$ and thus covers precisely the positive input set. During its execution, the algorithm maintains a priority queue of all candidate grammars that have been generated but not yet generalized. In each step, the algorithm removes the smallest grammar from the queue and applies to it all possible elementary generalizations, producing a new set of candidate grammars. Each resulting grammar is verified by our improved version of the Rekers-Schürr parser [8]. If a grammar covers at least one negative input graph, it is immediately discarded; otherwise, it is placed into the queue. To reduce the computational effort, only the *beamWidth* smallest grammars are kept in the queue. When the queue becomes empty, the algorithm outputs the smallest grammar created during the search process.

4.2   Elementary Generalizations

To perform a single step forward in our specific-to-general search, a given candidate grammar is 'slightly' generalized or merely restructured without changing its generative power. This is achieved by elementary generalizations of two types, called 'type A' and 'type B'.

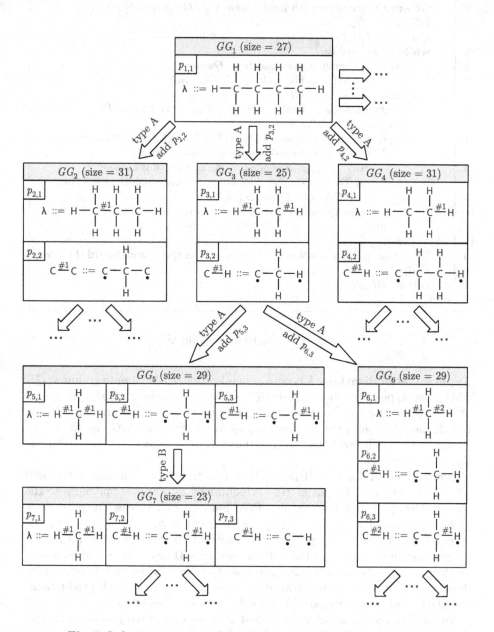

Fig. 2. Inducing a grammar from the structural formula of butane

```
1   procedure induceGrammar(𝒢⁺, 𝒢⁻, beamWidth, maxVertexCount)
2       GG₁ := the grammar with productions λ ::= G for each G ∈ 𝒢⁺;
3       GG_min := GG₁;
4       Queue := {GG₁};
5       while Queue ≠ ∅ do
6           GG := the smallest grammar from Queue;
7           if |GG| < |GG_min| then GG_min := GG end;
8           Queue := Queue \ {GG};
9           NewProductions := findProductions(GG, maxVertexCount);
10          foreach p ∈ NewProductions do
11              GG′ := type-A generalization of GG via the production p;
12              if GG′ does not cover any graph from 𝒢⁻ then
13                  GG″ := type-B generalization of GG′ (if it exists);
14                  if GG″ exists and does not cover any graph from 𝒢⁻ then
15                      Queue := Queue ∪ {GG″}
16                  else Queue := Queue ∪ {GG′} end
17                  end
18              end;
19              retain the beamWidth smallest grammars in Queue and discard the others
20          end;
21      return GG_min
22  end
```

Fig. 3. The induction algorithm

Type-A Generalization. A type-A generalization of a candidate grammar GG adds a new type-II production p to GG and reverse-applies it to all occurrences of $RHS(p)$ in the base graphs of GG, resulting in a new grammar GG'. In Fig. 2, the addition of the production $p_{3,2}$ to the grammar GG_1 results in the grammar GG_3. The grammar GG_3 is thus a type-A generalization of GG_1 via the production $p_{3,2}$.

How can we find a type-II production p to transform a grammar GG into GG'? Since the base graphs of GG' are obtained by reverse-applying p to the base graphs of GG, the base graphs of GG must contain at least one occurrence of $RHS(p)$, i.e., at least one subgraph of the form $[u(S)v]$. We thus search the base graphs of GG for all possible subgraphs of the form $[u(S)v]$. Each such subgraph is an occurrence of the RHS of some type-II production, and each such production is eligible to enrich the grammar GG. Since we cannot determine the 'best' type-II production in advance, we have to consider all such productions, and thus we obtain many possible type-A generalizations of GG.

To simplify the explanation, let us first focus on undirected graphs and grammars. The process of finding type-II productions to generalize an undirected grammar GG is outlined in Fig. 4. The auxiliary procedure findSubgraphs (omitted for lack of space) finds all subgraphs comprising up to $maxVertexCount$ vertices in the set of base graphs of GG. More precisely, the procedure creates a set of pairs (Subgraph, Occurrences), where Subgraph is a graph and Occurrences is a set of pairs $(Match, Host)$ such that $Match$ is an occurrence of the graph $Subgraph$ in the graph $Host \in \mathcal{B}(GG)$. The procedure findSubgraphs first finds all single-vertex subgraphs and then iteratively produces larger subgraphs as single-vertex extensions of individual occurrences of smaller subgraphs. This approach was inspired by a large selection of algorithms sharing similar goals [6], especially by the VSiGraM approach [14], which could be used in place of it.

```
1   procedure findProductions(GG, maxVertexCount)
2       SubsAndOccs := findSubgraphs(B(GG), maxVertexCount);
3       Productions := ∅;
4       foreach (Subgraph, Occurrences) ∈ SubsAndOccs do
5           foreach (Match, Host) ∈ Occurrences do
6               if |Nh_Host(Match)| = 2 then
7                   call the two vertices in Nh_Host(Match) u and v;
8                   A := label(u); B := label(v);
9                   I := {i | [A—#i— B] is a subgraph in GG};
10                  if I = ∅ then k := 1 else k := max(I) + 1 end;
11                  foreach i ∈ I ∪ {k} do
12                      Productions := Productions ∪ {[A—#i— B] ::= [A(Subgraph)B]}
13                  end
14              end
15          end
16      end;
17      return Productions
18  end;
```

Fig. 4. Finding eligible type-II productions to generalize an undirected grammar GG

After receiving a set of subgraph-occurrence pairs, the procedure findProductions searches this set for all subgraphs that can serve as possible production *cores* (not entire RHSs!). For a subgraph S to serve as the core of a production, S must have exactly two neighbors in the base graph in which it occurs. Such a subgraph S gives rise to a type-II production $p\colon [A \overset{\#i}{\text{—}} B] ::= [A(S)B]$, where $\#i$ could stand for any nonterminal label. If the graph $[A \overset{\#i}{\text{—}} B]$ already occurs as a subgraph somewhere in the grammar, then the production p actually generalizes the grammar GG; otherwise, GG is merely restructured. To take both possibilities into account, we create a separate production $[A \overset{\#i}{\text{—}} B] ::= [A(S)B]$ for each i such that $[A \overset{\#i}{\text{—}} B]$ occurs as a subgraph in GG and for a single value of i that does not meet this condition (lines 9–13 in Fig. 4). For example, the grammar GG_3 in Fig. 2 is extended by the productions $p_{5,3}$ (giving the grammar GG_5) and $p_{6,3}$ (giving the grammar GG_6), which differ only in their LHS edge labels.

In the case of directed graphs and grammars, a production core $[A(S)B]$ can serve as the RHS in two distinct production families, namely $[A \overset{\#i}{\longrightarrow} B] ::= [A(S)B]$ and $[B \overset{\#i}{\longrightarrow} A] ::= [B(S)A]$. Since neither of these families can be considered preferable in advance, both should be added to the resulting production set. The family

$[A \xrightarrow{\#i} B]$ ::= $[A(S)B]$ contains a production for each i such that $[A \xrightarrow{\#i} B]$ occurs as a subgraph in GG and for a single value of i that does not meet this condition. The other family is defined in an analogous fashion.

After each type-A generalization step, the resulting grammar is simplified by reverse-applying its type-II productions to its base graphs wherever possible and as many times as possible. This procedure is not essential for the induction process, but can make the grammar considerably smaller.

Type-B Generalization. A type-B generalization of a candidate grammar GG replaces two 'similar' productions of GG with a set of new productions, giving a grammar GG' that is at least as general as GG. In Fig. 2, the grammar GG_5 is generalized to GG_7 by replacing the productions $p_{5,2}$ and $p_{5,3}$ with $p_{7,2}$ and $p_{7,3}$. The notion of 'similar' productions is based on the concept of *unifiability*.

Edge labels l and m are *unifiable* (denoted $l \cong m$) if $(l = m) \vee (l \in \mathcal{N}^E) \vee (m \in \mathcal{N}^E)$. The *unification* of unifiable edge labels l and m (denoted $unif(l, m)$) is the label l if $l \in \mathcal{N}^E$; otherwise, $unif(l, m) = m$. Graphs G and H are *unifiable* if there exists a *unifying isomorphism* $h \colon G \to H$, i.e., a bijective vertex-to-vertex and edge-to-edge mapping such that $label(h(v)) = label(v)$, $conn(h(e)) = h(conn(e))$, and $label(h(e)) \cong label(e)$ for all $v \in V_G$ and $e \in E_G$. The *unification* of such graphs G and H (denoted $unif(G, H)$) is a graph obtained from G by setting $label(e) := unif(label(e), label(h(e)))$ for all edges $e \in G$.

Type-B generalization can be applied to a pair of directed productions p and q if they take the form $p \colon [u \colon A \xrightarrow{e \colon \#i} v \colon B]$ ::= $[u(S)v]$ and $q \colon [u' \colon A \xrightarrow{e' \colon \#i} v' \colon B]$ ::= $[u'(S')v']$ and if there exists a unifying isomorphism $h \colon RHS(p) \to RHS(q)$ such that $h(u) = u'$ and $h(v) = v'$. A type-B generalization step replaces the productions p and q with a set that comprises: (1) a production $[A \xrightarrow{\#i} B]$ ::= $[A(S'')B]$, where $S'' = unif(S, S')$ (let $g \colon S'' \to S$ and $g' \colon S'' \to S'$ denote the corresponding unifying isomorphisms); (2) a production $[P \xrightarrow{\#j} Q]$ ::= $[P \xrightarrow{r} Q]$ for each edge e of S'' such that $label(conn(e)) = (P, Q)$, $label(e) = \#j$, and $label(g(e)) = r \vee label(g'(e)) = r$. Undirected productions are treated in an analogous manner.

5   Experimental Results

5.1   Application to Flowcharts

In our first series of experiments, we applied the induction algorithm to various sets of valid flowchart graphs. Our goal was to induce a grammar that generates (a superset of) the language generated by the reference grammar in Fig. 1. We experimented with different sets of randomly generated flowchart graphs and different input parameters. Each input set comprised between 10 and 50 graphs with up to 25 vertices. Different sets gave rise to different grammars, but in many cases, the algorithm induced the grammar shown in Fig. 5 or some variation thereof. The productions p_1 through p_8 in Fig. 5 are equivalent to

the productions of the reference grammar despite the fact that all nonterminal
edges are reversed. (Note the positions of the markers 'S' and 'T'; in every type-
II production, the source vertex on the LHS coincides with the bottom vertex
on the RHS.) The induced grammar can therefore generate any valid flowchart.

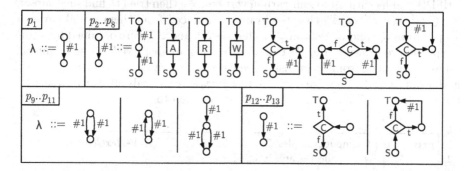

Fig. 5. A grammar induced from various sets of valid flowcharts

5.2 Application to Chemical Structural Formulas

In our second series of experiments, we tried to induce a grammar of linear hydrocarbons with single and double bonds (LHSDB). This graph language comprises
the structural formulas of chemical compounds consisting of carbon atoms (vertices labeled C) and hydrogen atoms (vertices labeled H). The carbon atoms
form a chain connected with single and double bonds (edges). The hydrogen
atoms are connected to the carbon atoms by means of single bonds so that every carbon atom has exactly four incident bonds. Some positive and negative
examples of the LHSDB language are shown in Fig. 8. Our reference grammar
for this language is depicted in Fig. 6.

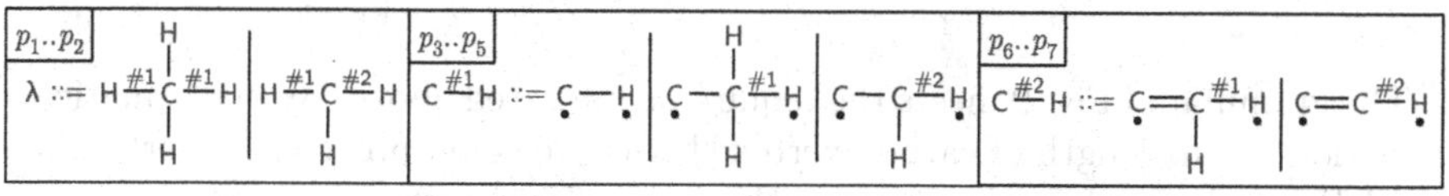

Fig. 6. The reference grammar for the LHSDB language

To make the induction problem more challenging, the induced grammar was
required to cover *only* valid (though not necessarily linear) hydrocarbons. We
thus demanded that the induced grammar cover all valid LHSDB graphs and
that every graph covered by the induced grammar represent a valid hydrocarbon.
Experiments showed that such a grammar probably cannot be induced from
positive graphs alone. Moreover, a correct grammar could not be induced from
'almost any' pair of input sets. A favorable combination of positive and negative
input graphs had to be sought more systematically.

To determine whether a correct LHSDB grammar can be induced and from what set of examples this can be achieved, we prepared a set of 42 positive examples $(\mathcal{G}_0^+)$ and a set of 200 negative examples $(\mathcal{G}_0^-)$. The positive set comprised all correct LHSDB graphs with up to 6 carbon vertices. The negative set was obtained by randomly removing one or two hydrogen atoms in correct LHSDB graphs with up to four carbon vertices. We then tried to find such subsets $\mathcal{S}^+ \subseteq \mathcal{G}_0^+$ and $\mathcal{S}^- \subseteq \mathcal{G}_0^-$ that the grammar induced from them would cover all graphs from $\mathcal{G}_0^+$ and none from $\mathcal{G}_0^-$. The sets $\mathcal{S}^+$ and $\mathcal{S}^-$ were obtained by a simple procedure shown in Fig. 7. The resulting sets are displayed in Fig. 8, and the grammar induced from them (using $beamWidth = 10$ and $maxVertexCount = 5$) is shown in Fig. 9. The size of the induced grammar equals 59. For comparison, the size of the reference grammar (Fig. 6) amounts to 54.

```
1   procedure findInputExamples(𝒢₀⁺, 𝒢₀⁻, beamWidth, maxVertexCount)
2       𝒮⁺ := {the smallest graph in 𝒢₀⁺};
3       𝒮⁻ := ∅;
4       GG := induceGrammar(𝒮⁺, 𝒮⁻, beamWidth, maxVertexCount);
5       Missed⁺ := {G ∈ 𝒢₀⁺ | GG does not cover G};
6       Missed⁻ := {G ∈ 𝒢₀⁻ | GG covers G};
7       while (Missed⁺ ≠ ∅) ∨ (Missed⁻ ≠ ∅) do
8           if Missed⁻ ≠ ∅ then 𝒮⁻ := 𝒮⁻ ∪ {the smallest graph from Missed⁻}
9           else 𝒮⁺ := 𝒮⁺ ∪ {the smallest graph from Missed⁺} end;
10          GG := induceGrammar(𝒮⁺, 𝒮⁻, beamWidth, maxVertexCount);
11          Missed⁺ := {G ∈ 𝒢₀⁺ | GG does not cover G};
12          Missed⁻ := {G ∈ 𝒢₀⁻ | GG covers G}
13      end;
14      return (𝒮⁺, 𝒮⁻)
15  end
```

Fig. 7. Extraction of a pair of small favorable input graph sets ($\mathcal{S}^+$ and $\mathcal{S}^-$) from a pair of larger disjoint graph sets ($\mathcal{G}_0^+$ and $\mathcal{G}_0^-$)

The grammar of Fig. 9 meets the requirements stated above. By mathematical induction on the length of carbon vertex chains, we could prove that every valid LHSDB graph can be generated by the induced grammar. To prove that the grammar generates only valid hydrocarbon graphs, we would have to show that every vertex introduced by the grammar eventually obtains the correct number of incident edges (four in the case of carbon vertices and one in the case of hydrogen vertices). To see this, consider that any subgraph $C\underline{}^{\#1}H$ expands into $C-(\ldots)-H$ and that any subgraph $C\underline{}^{\#2}H$ expands into $C=(\ldots)-H$.

Given the input sets $\mathcal{S}^+$ and $\mathcal{S}^-$ of Fig. 8, the induction algorithm was shown to be robust to the parameters $beamWidth$ and $maxVertexCount$, provided that $beamWidth \geq 1$ and $maxVertexCount \geq 3$. The values 1 and 2 for $maxVertexCount$ cannot possibly produce any meaningful results, since the production p_5 in Fig. 9, which seems to be an indispensable part of any valid

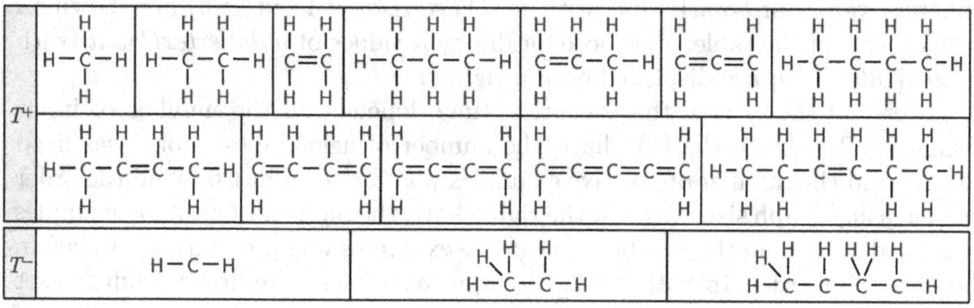

Fig. 8. The positive input set ($\mathcal{S}^+$) and the negative input set ($\mathcal{S}^-$) for the induction of an LHSDB grammar

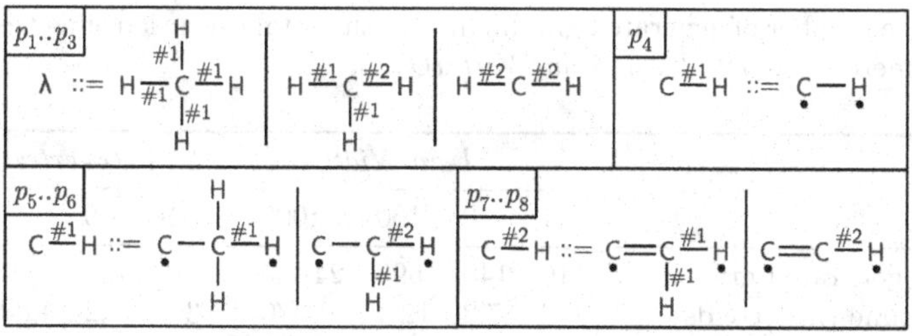

Fig. 9. The grammar induced from the input sets of Fig. 8

grammar, has three vertices in its core. We systematically varied both parameters and ran the induction algorithm for each pair of values. The resulting grammars were tested on a set of positive and negative graphs disjoint from $\mathcal{G}_0^+$ and $\mathcal{G}_0^-$.

5.3 Computational Complexity

Owing to the exhaustive subgraph enumeration procedure, which is the basis of type-A generalization, and to the Rekers-Schürr parser, our algorithm has exponential worst-case complexity in terms of time and memory consumption. The complexity of subgraph search could be reduced at the cost of missing some subgraphs. For example, the approaches of Jonyer et al. [11], Kukluk et al. [13], and Ates et al. [2] place an upper limit on the number of created subgraphs and hence run in a polynomial time and space at the cost of suboptimal results. The improved version of the Rekers-Schürr parser runs in polynomial time and space for many grammars [8], but its worst-case complexity is still exponential. This fact should not come as a surprise, since the problem of graph grammar parsing is NP-hard even for very restricted classes of grammars [19].

Table 1 shows the performance of our induction algorithm on the input graph set of Fig. 8 with respect to the parameters *beamWidth* and *maxVertexCount*. We measured the number of generated candidate grammars and the total execution time of the algorithm on an 1.86-GHz Intel Core 2 Duo machine. The results for different values of *beamWidth* (with *maxVertexCount* fixed at 5) are shown on the left side of the table, and those for different values of *maxVertexCount* (with *beamWidth* = 10) are displayed on the right side.

Figure 10 shows how the execution time depends on the number of input examples. To obtain the left chart, the number of negative examples was fixed at 200, and the number of positive examples was varied from 1 to 42 in the order of increasing graph size. To draw the right chart, the number of positive examples was fixed at 42, and the number of negative examples was varied from 1 to 200 in no particular order. In both cases, the input examples were drawn from the set of 42 positive and 200 negative examples that were supplied to the procedure of Fig. 7 when searching for a favorable input set for hydrocarbons. The parameters *beamWidth* and *maxVertexCount* were fixed at 10 and 5, respectively.

Table 1. The number of generated grammars and the total execution time with respect to the parameters *beamWidth* and *maxVertexCount*

	beamWidth				*maxVertexCount*			
	1	10	100	1000	3	5	7	9
Number of generated grammars	116	148	590	24 435	95	148	195	224
Execution time (in seconds)	6.8	7.2	12.2	370	3.4	7.2	12.4	17.5

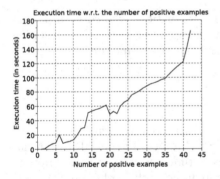
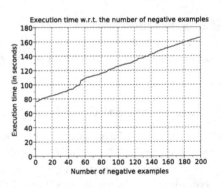

Fig. 10. Total execution time with respect to the number of input examples

The algorithm takes a little less than three minutes to finish if provided with the entire set of 42 positive and 200 negative examples. However, the user is not required to wait until the algorithm halts in order to obtain a meaningful grammar. Since the algorithm generates *only* grammars that are consistent with the input set, its result (the current minimum grammar) is valid at any point during its execution. The longer the algorithm runs, the smaller and the more general grammars it produces, but *all* induced grammars are valid with respect to the input set.

6   Conclusion

We have presented a novel graph grammar induction algorithm. Given a pair of disjoint graph sets, $\mathcal{G}^+$ and $\mathcal{G}^-$, the algorithm tries to find the smallest grammar that covers all graphs from $\mathcal{G}^+$ and none from $\mathcal{G}^-$. The induction process is realized as a parser-controlled specific-to-general search. We applied the proposed method to two meaningful and nontrivial graph languages. The algorithm exhibited a surprising inductive power when provided with favorable input.

In our 'chemical' example, a favorable input set was found by the algorithm in Fig. 7, which requires a pair of (large) initial input sets. To make the input selection process more 'user-friendly', we are working on a tool with the following interaction scenario: First, the user prepares a (small) set of positive input graphs. The tool induces a grammar from this set and generates a set of random graphs covered by the induced grammar. The user can then visually inspect the generated graphs and add to the negative input set all those that do not belong to the target language. If there are no such graphs, he or she may prepare some additional positive graphs and, by the help of the built-in parser, add to the positive input set all those graphs that are not covered by the induced grammar. After that, the tool induces a new grammar based on the updated input sets. The process repeats until the user is satisfied with the induced grammar.

At present, our research is focused on more general target grammar formalisms. In the formalism presented in this paper, a grammar for arbitrary hydrocarbons most probably does not exist. In the unrestricted LGG formalism, such a grammar comprises four simple productions (see the grammar GG_{HC} in Fig. 3 in [8]).

References

1. Aschenbrenner, N., Geiger, L.: Transforming Scene Graphs Using Triple Graph Grammars – A Practice Report. In: Schürr, A., Nagl, M., Zündorf, A. (eds.) AGTIVE 2007. LNCS, vol. 5088, pp. 32–43. Springer, Heidelberg (2008)
2. Ates, K., Kukluk, J.P., Holder, L.B., Cook, D.J., Zhang, K.: Graph grammar induction on structural data for visual programming. In: Proc. of the 18th IEEE International Conference on Tools with Artificial Intelligence, pp. 232–242. IEEE Computer Society, Washington, DC (2006)
3. Balogh, Z., Varró, D.: Model transformation by example using inductive logic programming. Software and Systems Modeling 8(3), 347–364 (2009)
4. Brijder, R., Blockeel, H.: On the inference of non-confluent NLC graph grammars. Journal of Logic and Computation (to appear, 2012)
5. Buchmann, T., Dotor, A., Uhrig, S., Westfechtel, B.: Model-Driven Software Development with Graph Transformations: A Comparative Case Study. In: Schürr, A., Nagl, M., Zündorf, A. (eds.) AGTIVE 2007. LNCS, vol. 5088, pp. 345–360. Springer, Heidelberg (2008)
6. Cook, D.J., Holder, L.B.: Mining Graph Data. John Wiley & Sons, New Jersey (2006)
7. Dubey, A., Jalote, P., Aggarwal, S.K.: Learning context-free grammar rules from a set of programs. IET Software 2(3), 223–240 (2008)

8. Fürst, L., Mernik, M., Mahnič, V.: Improving the graph grammar parser of Rekers and Schürr. IET Software 5(2), 246–261 (2011)
9. Javed, F., Mernik, M., Gray, J., Bryant, B.R.: MARS: A metamodel recovery system using grammar inference. Information and Software Technology 50(9-10), 948–968 (2008)
10. Jeltsch, E., Kreowski, H.J.: Grammatical Inference Based on Hyperedge Replacement. In: Ehrig, H., Kreowski, H.-J., Rozenberg, G. (eds.) Graph Grammars 1990. LNCS, vol. 532, pp. 461–474. Springer, Heidelberg (1991)
11. Jonyer, I., Holder, L.B., Cook, D.J.: MDL-based context-free graph grammar induction and applications. International Journal of Artificial Intelligence Tools 13(1), 65–79 (2004)
12. Kong, J., Zhang, K., Zeng, X.: Spatial graph grammars for graphical user interfaces. ACM Transactions on Computer–Human Interaction 13(2), 268–307 (2006)
13. Kukluk, J.P., Holder, L.B., Cook, D.J.: Inferring graph grammars by detecting overlap in frequent subgraphs. Applied Mathematics and Computer Science 18(2), 241–250 (2008)
14. Kuramochi, M., Karypis, G.: Finding frequent patterns in a large sparse graph. Data Mining and Knowledge Discovery 11(3), 243–271 (2005)
15. Nakamura, K., Matsumoto, M.: Incremental learning of context free grammars based on bottom-up parsing and search. Pattern Recognition 38(9), 1384–1392 (2005)
16. Parekh, R., Honavar, V.: Grammar inference, automata induction, and language acquisition. In: Dale, R., Somers, H.L., Moisl, H. (eds.) Handbook of Natural Language Processing, pp. 727–764. Marcel Dekker, New York (2000)
17. Plasmeijer, R., van Eekelen, M.: Term Graph Rewriting and Mobile Expressions in Functional Languages. In: Nagl, M., Schürr, A., Münch, M. (eds.) AGTIVE 1999. LNCS, vol. 1779, pp. 1–13. Springer, Heidelberg (2000)
18. Rekers, J., Schürr, A.: Defining and parsing visual languages with layered graph grammars. Journal of Visual Languages and Computing 8(1), 27–55 (1997)
19. Rozenberg, G. (ed.): Handbook of Graph Grammars and Computing by Graph Transformations, vol. 1: Foundations. World Scientific, River Edge (1997)
20. Schürr, A.: Specification of Graph Translators with Triple Graph Grammars. In: Mayr, E.W., Schmidt, G., Tinhofer, G. (eds.) WG 1994. LNCS, vol. 903, pp. 151–163. Springer, Heidelberg (1995)

Planning Self-adaption
with Graph Transformations

Matthias Tichy[1] and Benjamin Klöpper[2]

[1] Software Engineering Division,
Chalmers University of Technology and University of Gothenburg,
Gothenburg, Sweden
tichy@chalmers.se
[2] National Institute of Informatics (NII), Tokyo, Japan
kloepper@nii.co.jp

Abstract. Self-adaptive systems autonomously adjust their behavior in
order to achieve their goals despite changes in the environment and the
system itself. Self-adaption is typically implemented in software and
often expressed in terms of architectural reconfiguration. The graph transformation formalism is a natural way to model the architectural reconfiguration in self-adaptive systems. In this paper, we present (1) how we
employ graph transformations for the specification of architectural reconfiguration and (2) how we transform graph transformations into actions
of the Planning Domain Definition Language (PDDL) in order to use off-
the-shelf tools for the computation of self-adaptation plans. We illustrate
our approach by a self-healing process and show the results of a simulation
case study.

Keywords: Self-adaptive systems, graph transformations, planning,
PDDL.

1   Introduction

The complexity of today's systems enforces that more and more decisions are
taken by the system itself. This is resembled by the current trend to systems
which exhibit self-x properties like self-healing, self-optimizing, self-adaption.
Self-x properties cause additional complexity and dynamics within the system.
Therefore, appropriate development approaches have to be employed. The architecture is one of the key issues in building self-x systems [19,17]. In particular,
self-adaptation can be realized by adapting the architectural configuration by
adding and removing components as well as replacing them.

Kramer and Magee [17] presented a three-layer architecture for self-managed
systems consisting of the following layers: (1) *goal management*, (2) *change management*, and (3) *component control*. The component control layer contains the
architectural configuration of the self-adaptive system, i.e., the components and
their connections which are active in a certain state. Besides the execution of
the components, this layer is responsible for the execution of reconfiguration

A. Schürr, D. Varró, and G. Varró (Eds.): AGTIVE 2011, LNCS 7233, pp. 137–152, 2012.

plans. These plans, which describe the orderly adding, removing, and replacing of components and connectors, are executed to transform the current configuration into a new one in reaction to a new situation or event. They are stored in the change management layer and computed by the goal management layer.

Several approaches [25,18,28,4,26] employ graph transformations for the specification of architectural reconfiguration of self-adaptive systems. Graph transformations enable the application of formal verification approaches (e.g., [21,3]) and code generation [9] for execution during runtime by providing a sound formal basis. However, all of these approaches only address the modeling aspect of reconfiguration and do not address the computation of reconfiguration plans to meet the goals.

In this paper, we present how graph transformations can be integrated with automated planning approaches [12] to compute reconfiguration plans. We model the system structure using class diagrams and employ story patterns [9,29] as specific graph transformation formalism. Additionally, we extend story patterns by modeling elements for temporal properties to enable temporal planning. Similar to [6], we translate these models to the Planning Domain Definition Language [10] to enable the application of off-the-shelf planning software like SGPlan [5].

In the next section, we introduce the running example, which is about self-healing as a special case of self-adaptation, which is used to illustrate our approach. Section 3 gives a short introduction how we model the structure and the self-adaption behavior of our running example. The translation of the models to the planning domain definition language is described in section 4. Thereafter, we present an extension of our approach to durative actions and temporal planning in Section 5. In Section 6, we present results of simulation experiments for our self-healing application scenario. After a discussion of related work in Section 7, we conclude with an outlook on future work in Section 8.

2 Example

As an application example, we consider the self-healing process in an automotive application as presented in [16]. While it is currently not the case, we assume that, in the future, it is possible that software components can be deployed, started and stopped on electronic control units (ECU) at runtime. Currently, the deployment of software components to ECUs is done at design time but online reconfiguration gains interest and will eventually be realized. The AUTOSAR standard [11] with its standardized interfaces and the run-time environment (RTE) is the first step towards a system that can be reconfigured.

Our self-healing process reacts to failures of software components and hardware nodes (ECUs) by, for example, starting failed software components on working nodes, moving software components from a source to a target node, disconnecting and reconnecting software components. While the original self-healing process [16] considers redundant software components, we do not consider redundancy in this paper in order to keep the examples smaller.

Figure 1 shows an example of our self-healing process. On the left hand side of that figure, four nodes are shown which execute five component instances. node1 has experienced a failure and, thus, component c1 is not working anymore.

The self-healing process now reacts to this failure by computation and subsequently execution of a self-healing plan. This self-healing plan is comprised of the actions transfer which transfers the code of a component to a node, the action createInstance which instantiates the component of a node, and destroy which destroys a component instance on a node. For this example, the plan basically results in moving component c2 from node2 to node3 in order to free up space to subsequently instantiate the failed component c1 on node node2.

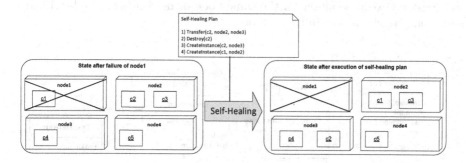

Fig. 1. Self-healing a failure of node1

A good example of a safety-relevant automotive subsystem is an adaptive cruise control (ACC). An ACC is an advanced tempomat, its functionality is to accelerate the car to the driver specified velocity, if no obstacle is detected. If an obstacle is detected, the car is first decelerated and then controls the gap between the car and the obstacle (mostly another car). The *adaptive* in its name comes from this change of behavior. Figure 2 shows the software components of a sample adaptive cruise control system. In this paper, we do not specifically target the self-healing of this adaptive cruise control system but address the general case of self-healing component-based systems as a running example.

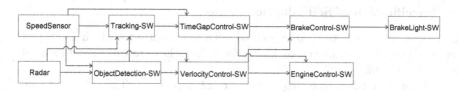

Fig. 2. Software components of an adaptive cruise control system [16]

3   Modeling with Graph Transformations

We employ the story pattern [9,29] graph transformation formalism, which is tightly integrated with the UML. It employs class diagrams for the specification of structure (similar to typed graphs in graph transformations) and refined collaboration diagrams for the specification of a graph transformation.

3.1   Specification of Structure

Figure 3 shows the class diagram for our self-healing scenario. Each component represents a software component. Pairs of components may have to communicate with each other. Nodes represent the computation hardware which are used to execute the software components. Before a component can be started on a node, the software code of the component has to be deployed to that specific node. Nodes are connected to other nodes. Components which communicate with each other must be executed either on the same node or on connected nodes. Each connection provides a certain transfer rate.

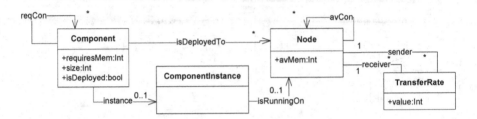

Fig. 3. Class diagram modeling the structure of the self-healing system

For the subsequent translation to PDDL, we require that for each class the maximum number of instances is specified by the developer. For our example, this maximum number of instances is known during design-time as the number of component types is known as well as the number of nodes. The number of component instances is equal to the number of component types as we specifically choose to have only a single instance of each component type in the system. Thus, this requirement is feasible for our scenario. This requirement is also typical in embedded systems. However, in other applications or domains this requirement might not hold.

3.2   Specification of Self-adaption Actions

The reconfiguration actions are specified with story patterns which are typed over the class diagram presented in Figure 3, i.e., they transform instances of this class diagram.

Story patterns follow the single pushout [22] formalism. The left hand side and the right hand side are merged into a single graph in story patterns. In this graph, all nodes and edges which are in the left hand side but not in the right hand side are marked as delete. These nodes and edges are deleted by the execution of the rule. All nodes and edges which are in the right hand side but not in the left hand side are marked as create. They are created by the execution of the rule. Simple negative application conditions can be modeled by appropriately annotating edges and nodes in the diagram. It is not allowed that negative nodes are attached to negative edges [29]. In contrast to standard story patterns, we do not support bound objects.

Fig. 4. Story pattern which specifies the deployment of a component c to node k

Figure 4 shows a story pattern which specifies the transfer of the code of a component c from a source node source to a target node k to enable starting the component on that node. The fact that the code of a component is available on a node is modeled as the existence of an isDeployedTo link between the component and the node. Additionally, the story pattern specifies that the amount of available memory on node k must be greater than the required memory of component c prior to execution. After execution the available memory is reduced.

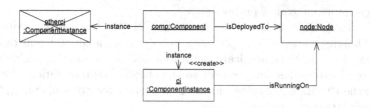

Fig. 5. Story pattern which specifies the instantiation of a component on a node

Figure 5 shows the story pattern concerning the instantiation of a component on a node. This story pattern can be executed if (1) the component's code has been previously deployed to the node and (2) the component is not already instantiated anywhere in the system. The first condition is expressed by the link isDeployedTo between comp and node. The second condition is expressed by the negative object otherci.

3.3 Specification of Goals

Finally, the goal has to be modeled for the self-healing system. Using only the left hand side of story patterns as in [8], it is possible to model a concrete situation which shall be reached by the computed plan. Though, this is rather cumbersome, especially for a large number of objects. Instead we use enhanced story patterns [24] which extend story patterns with subpatterns and quantifiers. Figure 6 shows the specification of the goal that for every component there should exist a component instance which is running on a node.

We do not provide modeling support for the initial state as the initial state is simply the current state of the self-adaptive system during runtime.

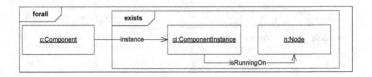

Fig. 6. Enhanced story pattern modeling the goal

4 Translation to PDDL

There exist several different representations for planning problems [12]. Set-theoretic approaches represent states by a set of propositions. Each action specifies which propositions have to hold in the state for the action to be applicable. Additionally, an action specifies which propositions will be added and removed to create a new state. The classical representation uses first-order literals and logical operators instead of propositions.

The Planning Domain Definition Language (PDDL) has been developed as a standard language for planning problems in order to compare different tools and algorithms. In order to be applicable for a wide variety of planning problems and tools, several extensions with respect to the classical representation have been added to the PDDL, e.g., typing, durative actions for temporal planning, fluents for representation of numerical values.

A planning problem in PDDL consists of two parts – the domain and the problem. The domain defines the types, the predicates, the functions, and the actions. A PDDL action is defined by a name, a list of parameters, a precondition and an effect. All objects which are referred to in the precondition and the effect have to be included in the list of parameters. The action can only be executed if the precondition is satisfied. The effect holds after the execution of the action. The problem defines the objects as well as the initial state and the goal state. The planning system then tries to compute a sequence of actions (including the arguments for the action parameters) which transforms the initial state into the goal state.

The PDDL extension :typing enables the modeling of types and generalization in the planning model. Thus, classes are translated to types in the domain definition; the associations are translated to predicates over the types. Story patterns are translated to actions in the planning domain. The left hand is translated to the precondition and the right hand side to the effect. Although the elements of both models map well to each other, there are details to consider.

4.1 Types, Predicates and Functions

Basically, classes are translated to types. Generalizations in the class diagrams are translated as well. Listing 1 shows the types generated from the class diagram of Figure 3. All types extend the general predefined type **object** denoted by the suffix "- object".

Listing 1: Mapping of classes to types in the PDDL

```
1 (:types
2    Component Node TransferRate ComponentInstance - object
3 )
```

Associations are translated to predicates over the source and target types. We support unidirectional and bidirectional associations. For example, the unidirectional association isDeployedTo between Component and Node is translated to the predicate (isDeployedTo ?component - Component ?node - Node). Listing 2 shows the predicates which are generated from the associations of the class diagram. For bidirectional associations, only one direction is translated to predicates in order to minimize the planning domain as bidirectional navigability is already provided by an unidirectional reference in PDDL.

Listing 2: Translation of associations to predicates

```
1  (:predicates
2     (exist ?object - object)
3     (isRunningOn ?component - Component ?node - Node)
4     (reqConn ?component - Component ?component - Component)
5     (instance ?component - Component ?componentinstance - ComponentInstance)
6     (isDeployedTo ?component - Component ?node - Node)
7     (avConnections ?node - Node ?node - Node)
8     (sourceTransferRate ?node - Node ?transferrate - TransferRate)
9     (targetTransferRate ?node - Node ?transferrate - TransferRate)
10    (runningOn ?componentinstance - ComponentInstance ?node - Node)
11    (isDeployed ?component - Component)
12 )
```

The PDDL prohibits the creation and deletion of objects to preserve a finite state space. As node creation and deletion is an important feature of graph transformations, we decided to emulate object creation and deletion by using the special predicate (exist ?object - object). We require a fixed number of objects of each type in the initial state of the planning problem. We emulate object creation and deletion by setting the exist predicate appropriately. Finally, Boolean attributes are also translated to predicates, e.g., the attribute isDeployed.

Functions provide mappings from object tuples to the realm of real numbers. This enables the translation of all numerical attributes from class diagrams. For example, the function (avMem ?n - Node) stores the amount of that node's memory which is increased and reduced depending on the number of components who are instantiated on a given node.

4.2 Actions

In the following, we present how story patterns are translated to PDDL actions. The preconditions and the effects of PDDL actions mirror naturally the left hand side and right hand side of a graph transformation. Listing 4 shows the PDDL action for the story pattern from Figure 4. We translate all nodes of the story

---

Listing 3: Translation of integer attributes to functions

```
1  (:functions
2    (requiresMem ?component - Component)
3    (size ?component - Component)
4    (avMem ?node - Node)
5    (value ?transferrate - TransferRate)
6  )
```

---

pattern to parameters of the action. The planner will bind these parameters to objects in such a way that the precondition is satisfied. The action transfer has three parameters for the two nodes source and k and the component c. The

---

Listing 4: Transfer of component code from source to target node

```
1  (:action transfer
2    :parameters (
3      ?source - Node ?k - Node ?c - Component
4    )
5    :precondition (and
6      (exist ?source) (exist ?k) (exist ?c)
7      (not (= ?source ?k))
8      (isDeployedTo ?c ?source)
9      (not (isDeployedTo ?c ?k))
10     (>= (avMem ?k) (requiresMem ?c))
11   )
12   :effect (and
13     (isDeployedTo ?c ?k)
14     (isDeployed ?c)
15     (decrease (avMem ?k) (requiresMem ?c))
16   )
17 )
```

---

precondition requires that all bound objects are indeed existing (line 6). The story pattern formalism uses a graph isomorphism, i.e., that two nodes of the graph transformation cannot be bound to the same node in the host graph. Consequently, we check in line 7 that the two nodes source and k are different. We translate the isDeployedTo-edge to the predicate in line 8.

The effect of the action simply states that the isDeployedTo predicate holds for the component c and the node k as the semantics of PDDL effects are that everything remains unchanged except the explicitly stated effect.

Attribute Expressions. Arithmetic expressions are supported by PDDL functions. The precondition can contain comparisons concerning functions. Values are assigned to functions in the effect. Concerning the transfer action, the precondition checks whether the available memory is greater or equal than the memory required by the component in line 10. In line 15, the available memory is decreased by this required amount of memory. Finally, the assignment of the Boolean variable isDeployed is part of the effect as well (line 14).

Negative Nodes and Edges. Story patterns enable the specification of simple negative application conditions by annotating nodes and edges that the node or the edge must not match.

The story pattern of Figure 4 contains a negative edge isDeployedTo between component c and node k. Thus, the story pattern is only applicable if the component c has not been already deployed to the node k. This is translated to a negated predicate as shown in line 9.

The case of a negative node is more complex. The semantics of a negative node [29] is that the matching of the left hand side minus the negative node must not be extendable to include a matching of the negative node as well. This is translated to a negative existential quantification over the objects of this type including the edges connected to the negative node as in lines 8 to 10 of Listing 5 which is the PDDL translation of the story pattern shown in Figure 5.

Again, special care has to be taken concerning injective matching. If a node is already positively matched, it will be excluded from the negative existential quantification.

Object Creation and Deletion. There exist several possibilities to emulate object creation and deletion. Naively, objects can be emulated by predicates which are set to true and false accordingly. This does only work for the case that it is not required that an object is identifiable. This is typically not suitable for graph transformations. As mentioned earlier, we decided to allocate a fixed number of objects of each class and use the additional predicate exist to denote whether the object exists or not.

Listing 5 shows the PDDL translation of Figure 5. Similar to Listing 4, we initially check for all nodes which are in the left hand side of the story pattern whether they exist in line 6. The object which will be created by the story pattern must not exist prior to the execution with the predicate in line 7. It is created in the effect (line 16).

Listing 5: Creating objects

```
1  (:action createInstance
2     :parameters (
3        ?comp - Component ?node - Node ?ci - ComponentInstance
4     )
5     :precondition (and
6        (exist ?comp) (exist ?node)
7        (not (exist ?ci))
8        (not (exists (?otherci - ComponentInstance)
9           (instance ?comp ?otherci)
10       ))
11       (not (instance ?comp ?ci)) (not (runningOn ?ci ?node))
12    )
13    :effect (and
14       (instance ?comp ?ci)
15       (runningOn ?ci ?node)
16       (exist ?ci)
17    )
18  )
```

Listing 6 shows how objects are destroyed by an action. The story pattern formalism follows the single pushout approach [22]. Therefore, we do not require that the dangling condition is satisfied and simply delete all edges related to the node (lines 6 and 7). The class diagram (see Figure 3) holds the information which edges we have to remove when destroying an object.

Listing 6: Destroying objects

```
1  (:action destroy
2   :parameters (?ci - ComponentInstance))
3   :precondition (and (exist ?ci))
4   :effect (and
5       (not (exist ?ci))
6       (forall (?o - Node) (not (runningOn ?ci ?o)))
7       (forall (?o - Component) (not (instance ?o ?ci)))
8   )
9  )
```

The model of the goal state shown in Figure 6 is translated to the PDDL in a similar way as the left hand side of story pattern with appropriate handling of the quantification.

5   Adding Temporal Properties

PDDL 2.1 [10] introduced syntax and semantics for temporal planning. Temporal planning relaxes the assumption of classical planning that events and actions have no duration. This abstraction is often not suitable as in reality actions *do* occur over a time span. Therefore, durations can be annotated to actions in temporal planning. As this allows concurrent actions, preconditions and effects have to be annotated. Three different temporal annotations are supported which can be combined: (1) at start, the precondition has to be satisfied at the beginning of the action, (2) at end, the precondition has to be satisfied at the end of the action, and (3), over all, the precondition has to be satisfied during the action. Effects have to be annotated with at start or at end.

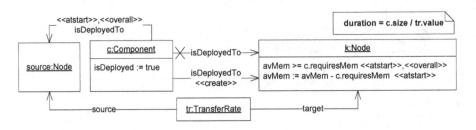

Fig. 7. Story pattern which specifies the deployment of a component c to node k including a duration

In general, story patterns do not consider time. Timed story patterns [14] are only concerned with *when* the pattern is executed, but not about the duration of its execution. We extend the story pattern by a duration fragment, which is used for the specification of the duration. Figure 7 shows this extension. The duration is computed based on the component size and the transfer rate for the connection between the nodes in our example. We annotate elements of the story pattern with the stereotypes ≪atstart≫, ≪atend≫, ≪overall≫ to specify the required temporal properties.

Listing 7: durative-action transfer

```
1   (:durative-action transfer
2     :parameters (?c - component ?k - Node ?source - Node ?tr - TransferRate)
3     :duration (= ?duration (/ (size ?c) (value ?tr)))
4     :condition (and
5       ...
6       (over all (isDeployedTo ?c ?source))
7       (at start (isDeployedTo ?c ?source))
8       (at start (sourceTransferRate ?rate ?source))
9       (at start (targetTransferRate ?rate ?target))
10      ...
11    )
12    :effect (and
13      (at end (isDeployedTo ?c ?k))
14      (at end (isDeployed ?c ))
15      (at start (decrease (avMem ?k) (requiresMem ?c)))
16    )
17  )
```

In contrast to the PDDL, we assume the following defaults in the case that the developer does not specify temporal stereotypes for the sake of visual clarity. All elements of the left hand side are assumed to have the stereotype ≪atstart≫ whereas all effects are assumed to have the stereotype ≪atend≫.

Listing 7 shows an excerpt from the durative action generated from the story pattern of Figure 7. The specification of the duration is shown in line 3. During the whole execution of the action, the component c must be deployed to the source node. As the temporal plan can schedule actions in parallel, we require that at the beginning of the action the available memory of node k must already be decreased by the required amount of component c.

6   Simulation Experiments

In order to show the feasibility of our approach, we conducted simulation experiments for the self-healing scenario. The scenario has been extended by resources which are required by components and provided by nodes, communication buses between the nodes as well as redundant allocation of components to nodes. The discrete event-based simulation environment simulates (1) failures of nodes, (2) repairs of nodes, and (3) periodic self-healing activities which are comprised of computing and executing self-healing plans. We abstract from the actual behavior of the components and restrict the simulation to the failures and the self-healing process. The plans are computed by the SGPlan automated planning software. The simulated system consists of 12 component types and 5 nodes connected by two communication buses. Node failures are randomly distributed by a negative exponential distribution $f_\lambda(x) = \lambda e^{-\lambda x}$ with a failure rate of $\lambda = 0.0001$. Every 5 time units, the self-healing part of the system checks whether any component type is not instantiated. In that case the planner is called and the resulting plan is executed.

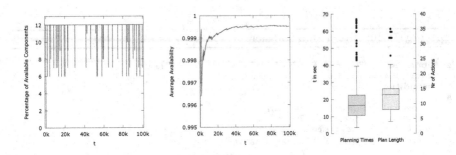

Fig. 8. Results from the simulation experiments

On the left side of Figure 8, the number of available component instances at each point of a single simulation run for 100.000 time units is shown. The complete system is available if each component type is instantiated on a node, i.e., if 12 component instances are available. At 53 points in time, a node fails which results in the failure of the components which are instantiated on that node. 10 of that 53 failures did happen to nodes which had no component types instantiated and, thus, resulted in no reduction of the number of component instances. After computing and executing the self-healing plan in reaction to a node failure, the number of available component instances increases to 12 again.

In the middle of Figure 8, the average availability of the system during the same simulation run is shown. We define availability as the probability that all component types of the system are instantiated at a certain point of time. The system starts with no instantiated components. Consequently, the average availability rises at the start of the simulation and reacts heavily to node failures. At the end of the simulation run, it is stable at 0.9995.

On the right side of that figure, we report the time taken by the planner to compute the plan as well as the planning length based on 277 calls to the planner. The planner is executed on an Intel Core2Duo with two cores at 2,53 GHz and 4 GB of RAM. Though, the planner uses only one core.

7 Related Work

In [4], a model-driven approach for self-adaption has been presented which also applied graph transformations for the specification of component reconfigurations. The graph transformations are used to specify goals, but the approach supports only the monitoring of these goals and not the computation of reconfiguration plans to achieve them. In general, planning is an important method in self-adapting and self-configuring systems. For instance, Arshad et al. [2] introduced a PDDL planning domain for automated deployment and reconfiguration of software in a distributed system. Satzger et al. [23] introduced a PDDL based planning approach for organic computing systems. Sykes et al. present in [13] an approach for planning architectural reconfiguration of component based systems based on the aforementioned three layer-architecture. They employ model checking as a planning technique based on labeled transition systems. This allows them to compute reactive plans, which generates actions sequence from every state in the state spaces towards the goal state.

None of the approaches supports the system developer appropriately in defining the required planning domains and offer techniques to check the correctness of the defined planning domain. A development process based on graph transformation naturally enables the support of respective modeling and verification tools and methods.

Vaquero et al. [27] present an approach for transforming UML models, use cases, class diagrams, state machines, and timing diagrams, to Petri Nets as well as PDDL in order to facilitate analysis and testing of requirements. The shown mapping from class diagrams to PDDL is similar to the one presented in this paper. In contrast to the approach by Vaquero et al., we use graph transformations for the specification of behavior which are more suitable for the specification of architectural reconfiguration.

There are only few tools and methodologies which support the designer in developing a planning model that complies to certain properties. For instance, Howey [15] et al. introduce VAL an automated tool that checks if plans generated by a planning system satisfy the specification made in the corresponding PDDL domain. Differently, PDVer is a tool that can be applied to check the correctness of Planning Domains [20]. However, PDVer does not formally verify the state transitions enabled by the planning domain, but heuristically generates and executes a number of test cases.

There are only a few approaches in the area of automated planning with graph transformations [6,8]. Edelkamp and Rensink present in [6] the combination of graph transformation and planning. They report that the employed planner (FF) can handle significantly bigger models than the graph transformation tool Groove. In contrast to our paper, Edelkamp and Rensink do not present how to automatically translate graph transformations to the input language of the employed planner. Estler [8] uses an A* as well as a Best First search for the computation of plans based on graph transformations. Instead of developing an own algorithm for planning, we employ standard off-the-shelf planning software which enables us to exploit their good performance and rich modeling properties, e.g., for temporal planning.

8   Conclusions and Future Work

We presented how graph transformation was used to specify actions for self-adaptive systems and how we use standard off-the-shelf automated planners to compute reconfiguration plans which order the execution of the reconfiguration actions. As a specific case of self-adaption we illustrated our approach by a self-healing process. We extended the employed story pattern formalism by several additional annotations for the specific case of durative actions in temporal planning. Based on this extensions, we showed how we translate story pattern to the Planning Domain Definition Language (PDDL) which is the standard planning language.

We have partially implemented the translation using the Eclipse Modeling Framework and Xpand as model-to-text translation environment. We used the EMF-based version of Fujaba, which is currently under development. We are currently working on finishing the implementation of the presented translation of attribute expressions as well as all syntax elements which are related to temporal planning.

Durative actions in the PDDL also include continuous effects which specify the continuous change of values during execution of the action, e.g., the physical position of an autonomous car based on its speed. It remains to be seen whether it makes sense to add those aspects to story patterns. Adding this might lead to a hybrid graph transformation formalism analogous to hybrid automata [1].

To reflect the specific strength and weaknesses of different planners as well as the differing requirements of application domains, it is reasonable to provide different translation schemes for story patterns to PDDL. The implementation and comparison of these different translations with respect to their effect on planners is an important part of our future research.

Story diagrams add control flow to story patterns. In order to use story diagrams for self-healing, we have to translate the control flow to PDDL as well. For the case of non-temporal planning, this works by numbering all story patterns and adding a sequence function which stores the current activity number. The control flow is then translated to appropriately handling this function in the precondition and the effect.

Acknowledgments. We thank Steffen Ziegert, Julian Suck, Florian Nafz, and Hella Seebach for discussions about the topic. We thank Christopher Gerking for the implementation of the prototypical translation of story patterns to PDDL as well as Alexander Stegmeier for the implementation of the simulation environment. Matthias Tichy was member of the software engineering group at the University of Paderborn, Germany, and the organic computing group at the University of Augsburg, Germany, while developing this approach. Benjamin Klöpper is a visiting researcher at NII and scholarship holder of the German Academic Exchange Service (DAAD).

References

1. Alur, R., Courcoubetis, C., Halbwachs, N., Henzinger, T.A., Ho, P.H., Nicollin, X., Olivero, A., Sifakis, J., Yovine, S.: The algorithmic analysis of hybrid systems. Theoretical Computer Science 138(1), 3–34 (1995)
2. Arshad, N., Heimbigner, D., Wolf, A.: Deployment and dynamic reconfiguration planning for distributed software systems. Software Quality Journal 15(3), 265–281 (2007)
3. Becker, B., Beyer, D., Giese, H., Klein, F., Schilling, D.: Symbolic invariant verification for systems with dynamic structural adaptation. In: Proc. of the 28th International Conference on Software Engineering, pp. 72–81. ACM Press (2006)
4. Becker, B., Giese, H.: Modeling of correct self-adaptive systems: A graph transformation system based approach. In: Proc. of the 5th International Conference on Soft Computing as Transdisciplinary Science and Technology, pp. 508–516. ACM, New York (2008)
5. Chen, Y., Wah, B.W., Hsu, C.W.: Temporal planning using subgoal partitioning and resolution in SGPlan. J. Artif. Intell. Research 26, 323–369 (2006)
6. Edelkamp, S., Rensink, A.: Graph transformation and AI planning. In: Edelkamp, S., Frank, J. (eds.) Knowledge Engineering Competition. Australian National University, Canberra (2007)
7. Ehrig, H., Engels, G., Kreowski, H.-J., Rozenberg, G. (eds.): TAGT 1998. LNCS, vol. 1764. Springer, Heidelberg (2000)
8. Estler, H.C., Wehrheim, H.: Heuristic search-based planning for graph transformation systems. In: Proc. of the Workshop on Knowledge Engineering for Planning and Scheduling (KEPS 2011), pp. 54–61 (2011)
9. Fischer, T., Niere, J., Torunski, L., Zündorf, A.: Story diagrams: A new graph rewrite language based on the Unified Modeling Language and Java. In: Ehrig et al. [7], pp. 296–309
10. Fox, M., Long, D.: PDDL2.1: An extension to PDDL for expressing temporal planning domains. J. Artif. Intell. Research 20, 61–124 (2003)
11. Fürst, S., Mössinger, J., Bunzel, S., Weber, T., Kirschke-Biller, F., Heitkämper, P., Kinkelin, G., Nishikawa, K., Lange, K.: AUTOSAR – A worldwide standard is on the road. In: Proc. of the 14th International VDI Congress Electronic Systems for Vehicles 2009. VDI (2009)
12. Ghallab, M., Nau, D., Traverso, P.: Automated Planning – Theory and Practice. Morgan Kaufmann Publishers (2004)
13. Heaven, W., Sykes, D., Magee, J., Kramer, J.: A Case Study in Goal-Driven Architectural Adaptation. In: Cheng, B.H.C., de Lemos, R., Giese, H., Inverardi, P., Magee, J. (eds.) Self-Adaptive Systems. LNCS, vol. 5525, pp. 109–127. Springer, Heidelberg (2009)
14. Heinzemann, C., Suck, J., Eckardt, T.: Reachability analysis on timed graph transformation systems. ECEASST 32 (2010)
15. Howey, R., Long, D., Fox, M.: VAL: Automatic plan validation, continuous effects and mixed initiative planning using PDDL. In: Proc. of the Int. Conference on Tools with Artificial Intelligence, pp. 294–301. IEEE Computer Society (2004)
16. Klöpper, B., Honiden, S., Meyer, J., Tichy, M.: Planning with utilities and state trajectories constraints for self-healing in automotive systems. In: Proc. of the 4th IEEE International Conference on Self-Adaptive and Self-Organizing Systems, pp. 74–83. IEEE Computer Society (2010)

17. Kramer, J., Magee, J.: Self-managed systems: An architectural challenge. In: Future of Software Engineering 2007, pp. 259–268. IEEE Computer Society (2007)
18. Le Métayer, D.: Describing software architecture styles using graph grammars. IEEE Transactions on Software Engineering 24(7), 521–533 (1998)
19. Oreizy, P., Medvidovic, N., Taylor, R.N.: Architecture-based runtime software evolution. In: Proc. of the 20th International Conference on Software Engineering, pp. 177–186. IEEE Computer Society (1998)
20. Raimondi, F., Pecheur, C., Brat, G.: PDVer, a tool to verify PDDL planning domains. In: Proc. of ICAPS 2009 Workshop on Verification and Validation of Planning and Scheduling Systems (2009)
21. Rensink, A.: The GROOVE Simulator: A Tool for State Space Generation. In: Pfaltz, J.L., Nagl, M., Böhlen, B. (eds.) AGTIVE 2003. LNCS, vol. 3062, pp. 479–485. Springer, Heidelberg (2004)
22. Rozenberg, G.: Handbook of Graph Grammars and Computing by Grah Transformation, vol. 1: Foundations. World Scientific (1997)
23. Satzger, B., Pietzowski, A., Trumler, W., Ungerer, T.: Using Automated Planning for Trusted Self-organising Organic Computing Systems. In: Rong, C., Jaatun, M.G., Sandnes, F.E., Yang, L.T., Ma, J. (eds.) ATC 2008. LNCS, vol. 5060, pp. 60–72. Springer, Heidelberg (2008)
24. Stallmann, F.: A Model-Driven Approach to Multi-Agent System Design. Ph.D. thesis, University of Paderborn (2009)
25. Taentzer, G., Goedicke, M., Meyer, T.: Dynamic change management by distributed graph transformation: Towards configurable distributed systems. In: Ehrig et al. [7], pp. 179–193
26. Tichy, M., Henkler, S., Holtmann, J., Oberthür, S.: Component story diagrams: A transformation language for component structures in mechatronic systems. In: Postproc. of the 4th Workshop on Object-oriented Modeling of Embedded Real-Time Systems. HNI Verlagsschriftenreihe (2008)
27. Vaquero, T.S., Silva, J.R., Ferreira, M., Tonidandel, F., Beck, J.C.: From requirements and analysis to PDDL in itSIMPLE3.0. In: Proc. of the International Competition on Knowledge Engineering for Planning and Scheduling (2009)
28. Wermelinger, M., Fiadeiro, J.L.: A graph transformation approach to software architecture reconfiguration. Sci. Computer Programming 44(2), 133–155 (2002)
29. Zündorf, A.: Rigorous Object Oriented Software Development. University of Paderborn (2002)

From Graph Transformation Units
via MiniSat to GrGen.NET

Marcus Ermler, Hans-Jörg Kreowski, Sabine Kuske, and Caroline von Totth*

University of Bremen, Department of Computer Science
PO Box 33 04 40, 28334 Bremen, Germany
{maermler,kreo,kuske,caro}@informatik.uni-bremen.de

Abstract. In logistics and other application areas, one faces many intractable and NP-hard problems like scheduling, routing, packing, planning, and various kinds of optimization. Many of them can nicely and correctly be modeled by means of graph transformation. But a graph transformation engine fails to run the solutions properly because it does not have any mechanisms to overcome or circumvent the intractability. In this paper, we propose to combine the graph transformation engine GrGen.NET with the SAT solver MiniSat to improve the situation. SAT solvers have proved to run efficiently in many cases in the area of chip design and verification. We want to take these positive experiences up and to use the SAT solver as a tentative experiment for assisting the graph transformation engine in finding solutions to NP-hard problems.

1 Introduction

Rule-based systems are nondeterministic in general because several rules may be applicable to some system state or one of the rules at several matches. In some cases, the nondeterminism is desired or harmless. A game would be extremely boring if there would be always a single step to perform (or none at all). A deterministic Chomsky grammar generates a single terminal word or the empty set and is more or less meaningless therefore. Sorting by exchanging two elements per step that are in the wrong order is nondeterministic, but yields always the same result whenever done as long as possible. Shortest paths, Eulerian cycles, minimum spanning trees or maximum flows can be computed nondeterministically in polynomial time with different results in general, but each result is acceptable. In many other cases, nondeterminism makes trouble. Even if each single run takes polynomial time, there may be an exponential number of runs for a given initial state and only a few of them may yield acceptable final states or none at all. Hence, if one runs such a rule-based system for some initial state once or for a small fixed number of attempts and no final state occurs, there may

---

* The authors would like to acknowledge that their research is partially supported by the Collaborative Research Centre 637 (Autonomous Cooperating Logistic Processes: A Paradigm Shift and Its Limitations) funded by the German Research Foundation (DFG).

A. Schürr, D. Varró, and G. Varró (Eds.): AGTIVE 2011, LNCS 7233, pp. 153–168, 2012.

be none or there may be one, but reachable by another run only. An exhaustive search is exponential, and any heuristic search may terminate fast, but fail to yield proper results in general. Many computational problems of this troublesome kind are of practical interest, and a good part of them can be found in the class of NP-hard problems (cf., e.g., [9]). For instance, many problems in logistics (and similar application domains) like scheduling, route planning and packing are NP-hard so that no efficient exact solutions are known and means are needed to overcome this intractability. Apart from using heuristics, another approach is carried out with some success in the area of chip design and verification: the use of SAT solvers (cf., e.g., [1,2,8]).

In this paper, we advocate the latter idea as an attempt to lessen the intractability of NP-hard problems in the area of graph transformation [18]. A graph-transformational specification of an NP-hard problem may be given by a graph transformation unit [11,12]. It specifies a set of derivations for each input graph which is of exponential size in general and only some of the derivations represent a proper solution of the problem (if it is solvable for the input at all). A graph transformation engine like GrGen.NET (see, e.g., [10]) can build up some derivations, but it has no means to guarantee success and the derived graphs may be far away from optimal solutions. To improve the situation, the graph transformation engine needs help.

We propose a prototypical system that combines GrGen.NET with the SAT solver MiniSat [5] and works as follows:

(1) A graph transformation unit of a special form is translated into a propositional formula in conjunctive normal form which describes all successful derivations.

(2) The formula is solved by the SAT solver yielding a variable assignment in the positive case that establishes one of the successful derivations.

(3) The variable assignment is translated into a control expression that guides the graph transformation engine to execute just the successful derivation and to visualize it in this way.

Altogether, this combination of GrGen.NET and MiniSat means that the graph transformation engine yields an exact solution of an NP-hard problem for some input whenever the SAT solver provides some positive result. To demonstrate the usefulness of our proposal, we discuss the job-shop scheduling problem, which is considered to be one of the most difficult scheduling problems in logistics (cf., e.g., [3,4,17]). It should be noted that the translation of graph transformation units into propositional formulas was already considered in [13] and is used in this paper with slight modifications. The interpretation of the variable assignments as successful derivations is new and very helpful because the intermediate representations as propositional formulas and variable assignments are very large (even for small input graphs) and very difficult to read. In contrast to this, the execution on GrGen.NET provides visualizations of the results.

The paper is organized as follows. In Sec. 2, the notion of graph transformation units is recalled. In Sec. 3, it is shown how the job-shop scheduling problem can be modeled as a graph transformation unit. In Sec. 4 graph transformation units are translated into propositional formulas so that they can be put into a SAT solver. Sec. 5 presents the prototypical implementation that combines the SAT solver MiniSat with the graph transformation engine GrGen.NET. Sec. 6 contains the conclusion.

2 Graph Transformation Units

Graph transformation units transform initial graphs to terminal graphs by applying rules according to control conditions. The components of graph transformation units are briefly recalled in the following.

Graphs. Graphs can be helpful to represent and visualize complex system states. Graph transformation units transform graphs and are independent of a concrete class of graphs. This means that transformed graphs belong to an underlying class $\mathcal{G}$ that can be chosen freely. For our running example, we use the class of *edge-labeled directed graphs without multiple edges*. For a set Σ of labels such a graph is a pair $G = (V, E)$ where V is a finite set of *nodes* and $E \subseteq V \times \Sigma \times V$ is a finite set of *labeled edges*. The components V and E are also denoted by V_G and E_G. An edge of the form (v, x, v) is a *loop*. In graph drawings we omit loops and write their labels inside their incident nodes.

The *empty graph* with no nodes and no edges is denoted by $\emptyset$. A graph G is a *subgraph* of a graph G' if $V_G \subseteq V_{G'}$ and $E_G \subseteq E_{G'}$. An *injective graph morphism* from a graph G to a graph H is an injective mapping $g \colon V_G \to V_H$ that preserves structure and labels, i.e., $(g(v), x, g(v')) \in E_H$ for each edge $(v, x, v') \in E_G$. The subgraph of H induced by g is denoted by $g(G)$.

Graph Class Expressions. A *graph class expression* is any expression that specifies a subset of the underlying graph class $\mathcal{G}$. A typical graph class expression is a type graph [6] which specifies the class of graphs of this type. Class diagrams can also be regarded as graph class expressions that specify all object diagrams covered by the class diagram. In this paper, we use concrete graphs as graph class expressions where each graph G specifies itself.

Rules. Rules are applied to graphs in order to transform them. Like the graph class, the rule class can be chosen freely as long as it is suitable to transform the graphs in $\mathcal{G}$. For our job-shop example, we use a rule class $\mathcal{R}$ that is similar to double-pushout rules. Concretely, such a *rule* r is a triple of graphs (L, K, R) where K is a subgraph of L and R. L is the *left-hand side*, K the *gluing graph* and R the *right-hand side*. We depict rules in the form $L \to R$ where the common nodes and edges are represented by using identical relative positions.

A rule $r = (L, K, R)$ is *applied* to a graph G according to the following steps. Choose an injective graph morphism g from L to G such that the removal of the items in $g(L) - g(K)$ from G does not produce dangling edges. Then remove the items in $g(L) - g(K)$ from G yielding D and add R disjointly to D. Afterwards, glue R and D as follows. (1) Merge each $v \in V_K$ with $g(v)$. (2) If there is an edge $(v, x, v') \in E_R - E_K$ with $v, v' \in V_K$ and an edge $(g(v), x, g(v')) \in E_D$ then these edges are identified.

For each $r \in \mathcal{R}$, the relation $\underset{r}{\Longrightarrow}$ denotes all pairs $(G, G') \in \mathcal{G} \times \mathcal{G}$ such that G can be transformed to G' by applying r. For each $P \subseteq \mathcal{R}$ the relation $\underset{P}{\Longrightarrow}$ is equal to $\bigcup_{r \in P} \underset{r}{\Longrightarrow}$ and $\underset{P}{\overset{*}{\Longrightarrow}}$ denotes the reflexive and transitive closure of $\underset{P}{\Longrightarrow}$. A sequence of rule applications is called a *derivation*.

Control Conditions. A *control condition* is any expression that specifies a set of derivations. Control conditions are employed to restrict the nondeterminism of graph transformation. In this paper, we use the following class $\mathcal{C}$ of control conditions. Every rule r is a control condition in $\mathcal{C}$ which requires one application of r, i.e., it permits the set of derivations $\{G \underset{r}{\Longrightarrow} G' \mid G, G' \in \mathcal{G}\}$. For every set $P \subseteq \mathcal{R}$, the expression $P!(k)$ is in $\mathcal{C}$ and prescribes to apply the rules of P as long as possible but at most k times. Finally, for each $c_1, c_2 \in \mathcal{C}$, the expressions $c_1; c_2$ and $c_1|c_2$ are in $\mathcal{C}$ where $c_1; c_2$ means to apply c_1 and afterwards c_2 and $c_1|c_2$ means to apply c_1 or c_2.

The introduced components give rise to the concept of graph transformation units defined as follows.

Graph Transformation Units. A *graph transformation unit* is a system $tu = (I, P, C, T)$ where I and T are graph class expressions called the *initial graph class expression* and the *terminal graph class expression*, $P \subseteq \mathcal{R}$, and $C \in \mathcal{C}$. The graph transformation unit tu specifies all pairs $(G, G') \in \underset{P}{\overset{*}{\Longrightarrow}}$ such that G is specified by I, G' is specified by T, and the underlying derivation is allowed by C. Such a derivation is called *successful*. In general graph transformation units can import other graph transformation unit, a feature not considered in this paper.

3   Modeling the Job-Shop Scheduling Problem

In this section, we model the job-shop scheduling problem as graph transformation unit.

3.1   The Job-Shop Scheduling Problem

The job-shop problem is a classical scheduling problem that consists of a finite set M of *machines*, a finite set $J \subseteq M^+$ of *jobs* where for each $j \in J$ and each $m \in M$ $count(m, j) \leq 1$, and a mapping $pt \colon T \to \mathbb{N}_{>0}$ with $T = \{(m, j) \in M \times J \mid count(m, j) = 1\}$.[1] The mapping pt associates a *processing time* with every *task* $t \in T$.

---

[1] M^+ denotes the set of nonempty words over M. For a symbol a and a word w, $count(a, w)$ counts the number of occurrences of a in w.

A *feasible schedule* is a mapping $start: T \to \mathbb{N}$ such that (1) for each job $j = m_1 \cdots m_k \in J$ with $k > 1$, $start(m_i, j) + pt(m_i, j) \leq start(m_{i+1}, j)$, for $i = 1, \ldots k - 1$; and (2) for each $m \in M$ and all $j, j' \in J$ with $j \neq j'$, $start(m, j) + pt(m, j) \leq start(m, j')$ or $start(m, j') + pt(m, j') \leq start(m, j)$. The first property means that the machines of each job j must be visited sequentially in the order in which they occur in j. The second property assures that each machine can process at most one job at the same time.

The *makespan* of each schedule *start* is defined as the maximum number in $\{start(t) + pt(t) \mid t \in T\}$. A *best schedule* is one where makespan is minimized. If a deadline is given, an interesting question is whether there exists a schedule at all the makespan of which meets the deadline. This *decision problem variant* of the job-shop scheduling problem is NP-complete [9].

In the following we do not allow interruptions in schedules, i.e., all machines perform a processing step if possible.

3.2   The Graph Transformation Unit *jobshop*

Let $JSP = (M, J, pt)$ be a job-shop scheduling problem and let $l \in \mathbb{N}$. Then we can construct a graph transformation unit $jobshop(M, J, pt, l)$ such that the semantics of $jobshop(M, J, pt, l)$ is not empty if and only if there is a feasible schedule the makespan of which is at most l. The components of $jobshop(M, J, pt, l)$ are the following.

The initial graph contains a node with an *idle*-loop for each machine and a node for each task. Moreover, there is a node with a *makespan*-loop and a 0-loop. For each job $j = m_1 \cdots m_k$, there is a *next*-edge pointing from task node (m_i, j) to the task node (m_{i+1}, j) $(i = 1, \ldots, k-1)$. There is also an edge from (m_i, j) to the machine node m_i labeled with $pt(m_i, j)$. Moreover, the first task node (m_1, j) is connected to the machine node m_1 via a *waiting*-edge. An example of an initial graph with three machines and three jobs generated by GrGen.NET is given in Fig. 1. For a better readability, every task-node of this graph additionally shows a job number and the position where it occurs within the job. For example, the two leftmost lower task nodes $t11$ and $t12$ are associated with job 1 consisting of the machine sequence $m1, m2$ as indicated by the positions, the *next*-edges and the *pt*-pointers originating from these task nodes. The first task $t11$ is waiting at machine $m1$ and it needs 2 time units to be processed.

The graph transformation unit $jobshop(M, J, pt, l)$ contains the rules *select*, *process*, *switch*, *finish*, *activate*, and *count*. The first two rules are given in Fig. 2. The rule *select* exchanges a *waiting*-edge from a task to an idle machine for a *selected*-edge from the machine to the task and makes the machine active. The rule *process* decreases the residual processing time of a task which is selected by an active machine by 1 and exchanges the *active*-loop at the machine for a *ready*-loop. As indicated below the arrow, this rule can only be applied if the remaining processing time has not reached 0.

The rules *switch* and *finish* are given in Fig. 3. The rule *switch* is applied when a task is processed. In this case the corresponding *selected*-edge and the 0-edge are removed. The machine which had selected the task becomes idle and

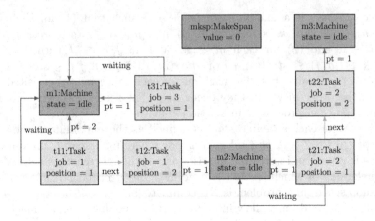

Fig. 1. An initial graph for job-shop

the next task of the same job is attached via a *waiting*-edge to its machine. The rule *finish* is applied when all tasks of the same job are processed. It removes the *selected*-edge and the 0-edge and changes the *ready*-loop of the machine to an *idle*-loop.

The rule *activate* in Fig. 4 relabels a *ready*-loop by an *active*-loop and the rule *count* increases the counter of the makespan node by one.

The control condition of $jobshop(M, J, pt, l)$ is

$$(select!; process!; switch!; finish!; activate!; count)^l$$

where for $r \in \{select, process, switch, finish, activate\}$ the expression $r!$ abbreviates the expression $\{r\}!(|M|)$ where $|M|$ denotes the number of machines. The control condition makes sure that in every iteration of *select!* ; *process!* ; *switch!* ; *finish!* ; *activate!* ; *count* the following happens. (1) Every idle machine with a waiting task becomes active by selecting the task (*select!*). (2) Every active machine becomes ready by decreasing the residual processing time of its selected task by 1 (*process!*). (3) If a task (m, j) with a next task (m', j) is processed, it becomes an isolated node, m becomes idle, and (m', j) starts to wait at machine m'. (*switch!*). (4) If a task (m, j) is processed and has no next task, it becomes an isolated node and m becomes idle (*finish!*). (5) Every ready machine is

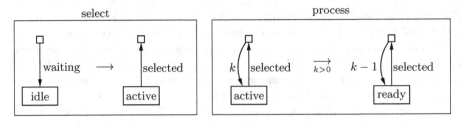

Fig. 2. The graph transformation rules *select* and *process*

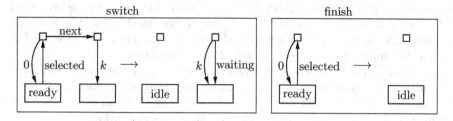

Fig. 3. The graph transformation rules *switch* and *finish*

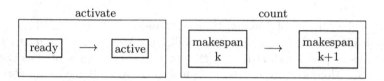

Fig. 4. The graph transformation rules *activate* and *count*

activated for the next processing step (*activate!*). (6) The makespan is increased by 1 (*count*).

The terminal graph class expression of *jobshop*(M, J, pt, l) requires that after executing the control condition all tasks should be finished. Concretely, the terminal graph consists of a node for every machine and every task as well as a node with a *makespan*-loop and an *l*-loop.

The following observation states the correctness of the presented unit. The proof is roughly sketched, a complete proof is beyond the scope of the paper.

Observation 1. Let $G, G' \in \mathcal{G}$, let (M, J, pt) be a job shop scheduling problem and let $l \in \mathbb{N}$. Then (G, G') is specified by *jobshop*(M, J, pt, l) if and only if there is a feasible schedule *start* of (M, J, pt) such that the makespan of *start* is at most l.

Proof (sketch). Let (G, G') be specified by *jobshop*(M, J, pt, l) and let *der* be a derivation from G to G', which is permitted by the control condition of *jobshop*(M, J, pt, l). Then the following statements hold by induction:

1. The rule *select* is applied in *der* exactly once to every task in G.
2. In every graph of *der*, each machine has at most one selected task.
3. In each execution of $c = select!; process!; switch!; finish!; activate!; count$ a machine can select at most one task.
4. In each execution of c, every selected task is processed one time unit.
5. Each selected task remains selected until it is completely processed.
6. A target of a *next*-edge in G can only be selected in *der* if its source is completely processed.
7. A task becomes an isolated node if and only if it is completely processed.

Choose $start(t) = i$, if *select* is applied to task t in the ith execution of c. Statements 2 to 5 imply that each machine must process a task completely before selecting a new one. Statement 4 to 6 imply that all tasks are processed in the order in which they occur in the jobs. Hence, $start$ is feasible. Since G' contains only isolated tasks we get by Statement 7 that all tasks are processed in G' and this implies together with 4 and 5 that $l \geq \max\{start(t) + pt(t) \mid t \in T\}$.

Conversely, let $start$ be a feasible schedule (without interruptions) of (M, J, pt) with makespan l and let $t_1, \ldots, t_n$ be an arrangement of the tasks in T such that $start(t_i) \leq start(t_{i+1})$ for $i = 1, \ldots, n - 1$. Let G be an initial graph of $jobshop(M, J, pt, l)$. Then by induction on n there is a derivation

$$G = G_0 \underset{P}{\overset{*}{\Rightarrow}} G_1 \underset{P}{\overset{*}{\Rightarrow}} \cdots \underset{P}{\overset{*}{\Rightarrow}} G_n \underset{P}{\overset{*}{\Rightarrow}} G'$$

where P is the rule set of $jobshop(M, J, pt, l)$ and for $i = 0, \ldots n - 1$ the graph G_{i+1} is obtained from G_i by applying condition c $start(t_{i+1}) - start(t_i)$ times and all tasks in $\{t \in T \mid start(t) + pt(t) \leq start(t_i)\}$ are represented as isolated nodes in G_i. Moreover, G' is obtained from G_n by applying c $l - start(t_n)$ times with the effect that all non-isolated tasks of G_n become isolated and the makespan counter (which is in G_n equal to $start(t_n)$) becomes equal to l. Hence, G' is a terminal graph that is obtained from G by applying condition c l times. Altogether we get that (G, G') is specified by $jobshop(M, J, pt, l)$.

4   From Graph Transformation Units to Propositional Formulas

In this section, we shortly describe how graph transformation units can be translated into propositional formulas in such a way that a SAT solver can be used to decide whether there is a successful derivation in the unit. In the case of the job-shop scenario, this means that the SAT solver decides whether there is a makespan of length at most l. We recall the constructions and results of [13] and illustrate them with small examples concerning the job shop scheduling problem. Additionally, we extend the class of control conditions used in [13] by the operators | and !(k) presented in Sec. 2.

Representing Graph Sequences as Propositional Formulas. We assume that the nodes of each graph G are numbered from 1 to n, i.e, $V_G = \{1, \ldots, n\} = [n]$, and we call n the *size* of G. Every propositional formula with variable set $\{edge(e, i) \mid e \in [n] \times \Sigma \times [n], i \in [m]\}$ represents a sequence $G_1, \ldots, G_m$ of graphs for each variable assignment f satisfying the formula. In more detail, graph G_i contains the edge e if and only if $f(edge(e, i)) = true$.

The graphs of the rule *select* in Fig. 2 without the *idle*-loop and the *active*-loop can be regarded as a small example of a graph sequence G_0, G_1 where G_0 is the left-hand side of *select*, G_1 the right-hand side, the upper node has number 1 and the lower number 2. Hence, by choosing $\Sigma = \{waiting, selected\}$, this sequence can be represented by the formula

$$edge(1, waiting, 2, 0) \wedge edge(2, selected, 1, 1) \wedge$$
$$\neg edge(1, waiting, 1, 0) \wedge \neg edge(2, selected, 2, 1) \wedge$$
$$\bigwedge_{i,j \in [2]} (\neg edge(i, selected, j, 0) \wedge \neg edge(i, waiting, j, 1)) \wedge$$
$$\bigwedge_{i \in [2]} (\neg edge(2, waiting, i, 0) \wedge \neg edge(1, selected, i, 1)).$$

Rule Applications as Propositional Formulas. For technical simplicity, we assume that the nodes of all three graphs in a rule coincide, i.e., for each $(L, K, R) \in \mathcal{R}$ the sizes of L, K, and R are equal. For a rule $r = (L, K, R)$ the set of injective mappings from $[size(L)]$ to $[n]$ is denoted by $\mathcal{M}(r, n)$. Let $r \in \mathcal{R}$, let $g \in \mathcal{M}(r, n)$ and let $i \in \mathbb{N}$. Then the formula

$$apply(r, g, i) = morph(r, g, i) \wedge rem(r, g, i) \wedge add(r, g, i) \wedge keep(r, g, i)$$

models the application of rule r to the graph $([n], E_i)$ with respect to morphism g, where

- $morph(r, g, i) = \bigwedge_{(v,x,v') \in E_L} edge(g(v), x, g(v'), i)$
- $rem(r, g, i) = \bigwedge_{(v,x,v') \in E_L - E_R} \neg edge(g(v), x, g(v'), i + 1)$
- $add(r, g, i) = \bigwedge_{(v,x,v') \in E_R} edge(g(v), x, g(v'), i + 1)$
- $keep(r, g, i) = \bigwedge_{(v,x,v') \notin g(E_L \cup E_R)} (edge(v, x, v', i) \leftrightarrow edge(v, x, v', i + 1)).$[2]

The formula $morph$ checks whether g is a graph morphism. The formula rem models the removal of all edges in $g(L) - g(R)$ and add models the addition of every edge in E_R which does not have an image under g. The formula $keep$ prescribes that all edges not in $g(L)$ or $g(R)$ should be present in the resulting graph $([n], E_{i+1})$ if and only if they are contained in $([n], E_i)$.

For example, let G_0 be the initial graph in Fig. 1 and let g be an injective mapping from the node set $\{1, 2\}$ of the rule $select$ (where 1 is the upper node) to the node set of G_0. Then the formula

$$morph(select, g, 0) \wedge rem(select, g, 0) \wedge add(select, g, 0)$$

is equal to

$$edge(g(1), waiting, g(2), 0) \wedge edge(g(2), idle, g(2), 0) \wedge$$
$$\neg edge(g(1), waiting, g(2), 1) \wedge \neg edge(g(2), idle, g(2), 1) \wedge$$
$$edge(g(2), selected, g(1), 1) \wedge edge(g(2), active, g(2), 1).$$

The formula $keep(select, g, 0)$ contains the term $edge(e, 0) \leftrightarrow edge(e, 1)$ for all variables $e \in ([9] \times \Sigma_{(M,J,pt,l)} \times [9]) \setminus \{(g(1), waiting, g(2)), (g(2), idle, g(2)), (g(2), selected, g(1)), (g(2), active, g(2))\}$ where (M, J, pt) is the modeled job shop scheduling problem and $\Sigma_{(M,J,pt,l)}$ is equal to $\{waiting, selected, next, idle, active, ready, makespan\} \cup \{pt(t) \mid t \in T\} \cup \{0, \ldots, l\}$, for each $l \in \mathbb{N}$.

---

[2] $g(E_L \cup E_R)$ abbreviates the set $\{(g(v), x, g(v')) \mid (v, x, v') \in E_L \cup E_R\}$.

Translation of Control Conditions into Propositional Formulas. In order to be able to generate propositional formulas from control conditions we associate a length len with every condition such that $len(r) = 1$, $len(P!(k)) = k$, $len(c_1; c_2) = len(c_1) + len(c_2)$, and $len(c_1|c_2) = \max\{len(c_1), len(c_2)\}$.

For example, for $c = select!; process!; switch!; finish!; activate!; count$ and $l \in \mathbb{N}$, we have $len(c^l) = l \cdot len(c) = l \cdot (4 \cdot |M| + 1)$ where $|M|$ is the number of machines (remember that here $r!$ stands for $\{r\}!(|M|)$). Hence, the length of the control condition of the *jobshop*-unit is equal to $l \cdot (4 \cdot |M| + 1)$.

Moreover, for each control condition c and each $i \in \mathbb{N}$ we denote the *formula of c at position i w.r.t. graph size n* as $f_n(c, i)$ defined as follows.

- $f_n(r, i) = \bigvee_{g \in \mathcal{M}(r,n)} apply(r, g, i)$, i.e., $f_n(r, i)$ specifies the application of r to graph $G_i = ([n], E_i)$.

- $f_n(P!(k), i) = \bigwedge_{j=i}^{i+k-1} \Big(\bigvee_{r \in P, g \in \mathcal{M}(r,n)} (morph(r, g, j) \wedge apply(r, g, j)) \vee$
  $\bigwedge_{r \in P, g \in \mathcal{M}(r,n)} \neg morph(r, g, j) \wedge apply(\emptyset, \emptyset, j) \Big)$

  This means that for $j = i$ to $i + k - 1$ some rule of P is applied to G_j if possible. If there is none of the rules in P applicable to G_j, i.e., if $\bigwedge_{r \in P, g \in \mathcal{M}(r,n)} \neg morph(r, g, j)$, the empty rule $\emptyset = (\emptyset, \emptyset, \emptyset)$ is applied, which has no effect.

- $f_n(c_1; c_2, i) = f_n(c_1, i) \wedge f_n(c_2, i + len(c_1))$, i.e., $f_n(c_1; c_2, i)$ applies c_1 of graph G_i and c_2 to graph $G_{i+len(c_1)}$.

- $f_n(c_1|c_2, i) = f_n(c_1; \emptyset^{len(c_1|c_2)-len(c_1)}, i) \vee f_n(c_2, ; \emptyset^{len(c_1|c_2)-len(c_2)}, i)$. This means that $f_n(c_1|c_2, i)$ applies c_1 or c_2 but if the length of the applied condition c_i is shorter than the length of $c_1|c_2$, then the empty rule is applied $len(c_1|c_2) - len(c_i)$ times.

The formula for a control condition C in a graph transformation unit is then equal to $f_n(C, 0)$ for each graph size n.

For example, if $c = select!; process!; switch!; finish!; activate!; count$, we have $f_n(c^l, 0) = f_n(c, 0) \wedge f_n(c, 1 \cdot len(c)) \wedge \cdots \wedge f_n(c, l \cdot len(c))$, and for $i = 0, \ldots, l$, $f_n(c, i \cdot len(c))$ is equal to

$$f_n(select!, i \cdot len(c)) \wedge f_n(process!; switch!; finish!; activate!; count), i \cdot len(c) + |M|).$$

Graph Class Expressions as Propositional Formulas. Every initial graph I can by represented by the formula

$$prop(I, 0) = \bigwedge_{e \in E_I} edge(e, 0) \wedge \bigwedge_{e \notin E_I} \neg edge(e, 0).$$

Similarly, any terminal graph T can be encoded as the formula $prop(T, len(C))$, which is obtained from $prop(I, 0)$ by replacing I by T and 0 by $len(C)$ where C is the control condition of the graph transformation unit. To illustrate this with a small graph, we take as I the left-hand side of the rule *select* (without the *idle*-loop) and translate it into the corresponding formula $prop(I, 0)$ similarly to the first example of this section which translated the left as well as the right-hand side of *select*. For $\Sigma = \{waiting\}$, the result is equal to

$$edge(1, waiting, 2, 0) \wedge \neg edge(1, waiting, 1, 0) \wedge \bigwedge_{i \in [2]}(\neg edge(2, waiting, i, 0)).$$

More general examples of graph class expressions are considered in [13].

Putting the described formulas together, we can express graph transformation units as propositional formulas as follows.

Graph Transformation Units as Propositional Formulas. Every graph transformation unit $tu = (I, P, C, T)$ can be encoded as the propositional formula

$$fder(tu) = prop(I, 0) \wedge f_n(C) \wedge prop(T, len(C))$$

where n is the size of I.

As stated in [13], every variable assignment that satisfies the formula specifies a sequence of graphs corresponding to a successful derivation in tu.

It is worth noting that the employment of a SAT solver only makes sense if the number of literals in the resulting formula is bounded by a polynomial. For graph transformation units in which each allowed derivation is of polynomial length this is the case (cf. [13]). Fortunately, all derivations of the transformation unit $jobshop(M, J, pt, l)$ consist of $l \cdot (4 \cdot |M| + 1)$ rule applications, which implies that they have polynomial lengths.

As shown in the next section one can extract a derivation from every satisfying variable assignment.

5   Combining MiniSat and GrGen.NET

In this section we take a closer look at a prototypical implementation in Haskell which combines the SAT solver MiniSat with the graph transformation engine GrGen.NET. In reference to its task the tool has been called SATaGraT which is the abbreviation for *SAT solver assists graph transformation engine*. SATaGraT uses MiniSat 2, GrGen.NET 3.0, and Glasgow Haskell Compiler 6.12 and has been tested on an Intel 3.2 GHz with 8GB RAM.

5.1   System Description

Fig. 5 shows an overview of the structure of the implemented system which is explained in detail in the following. SATaGraT takes a graph transformation unit as given in Sec. 2 as input and generates a propositional formula as defined in Sec. 4 which describes all possible runs of this unit. In order to obtain a satisfying variable assignment, the generated propositional formula is translated into conjunctive normal form and fed into MiniSat, a fast and easy-to-use SAT solver. The satisfiability preserving Tseitin transformation [20] which allows a

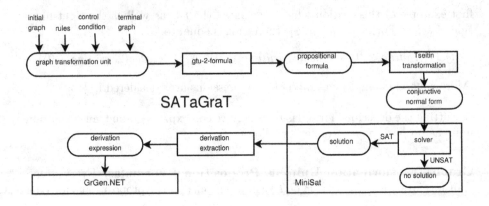

Fig. 5. The structure of SATaGraT

conversion with linear time and space requirements is used for the translation into conjunctive normal form (cnf for short). In cases in which the formula is satisfiable, MiniSat delivers a satisfiying assignment of the variables which is then used to construct a successful derivation of the initial graph in the original graph transformation unit.

MiniSat is fully integrated in SATaGraT, i.e., the generated conjunctive normal form is saved in a file in the DIMACS format as input for MiniSat, then MiniSat solves the conjunctive normal form, and at last SATaGraT reads the output from MiniSat and saves the assignment of the variables in an internal representation. MiniSat returns a list of variable assignments which includes the information about the solution as well as all the extra variables of the Tseitin transformation. In Fig. 6, one can see a snippet of the MiniSat-output with the variable numbers 41820 to 41878 where the truth value of variables preceded by a minus sign is false and the truth value of unsigned variables is true. Such an output is obviously cryptic and hard to interpret. Therefore, a suitable visualization seems quite useful.

```
SAT
...
41820 -41821 -41822 -41823 -41824 -41825 -41826 -41827 -41828 -41829 -41830 -41831 -41832 -41833
-41834 -41835 -41836 -41837 -41838 -41839 -41840 -41841 -41842 -41843 -41844 -41845 -41846 41847
41848 41849 41850 41851 41852 41853 41854 41855 41856 41857 41858 41859 41860 41861 41862 41863
41864 41865 41866 41867 41868 41869 41870 41871 41872 41873 41874 41875 41876 41877 41878
...
```

Fig. 6. A snippet of the output of MiniSat

In accordance with Sec. 4, the assignment of the variables yields a sequence of graphs representing the derivation. These graphs are described by their edges as the set of nodes remains invariant throughout the derivation process. In order to use GrGen.NET for constructing and running the derivation, the applied rules and the matchings used have to be extracted from this sequence. The following program in pseudocode illustrates the idea behind the implemented algorithm for extracting rules and matchings from the variable assignment, where $g(E_L) = \{(g(v), a, g(v')) \mid (v, a, v') \in E_L\}$ and $g(E_R) = \{(g(v), a, g(v')) \mid (v, a, v') \in E_R\}$ for every rule $(L, K, R) \in P$ and every injective mapping $g \in \mathcal{M}(r, n)$:

FUNCTION ExtractRulesAndMatchings $(E_0, \ldots, E_{derivLength} : SetOfEdges,$
$\qquad\qquad P : SetOfRules, \ n : |V|, \ derivLength : DerivationLength)$
$\qquad\qquad : (Derivation, \ SetOfEdges)$
$\quad$ FOR $m = 1, \ldots, derivLength, (L, K, R) \in P, g \in \mathcal{M}((L, K, R), n)$ DO
$\qquad \overline{E}_m := E_m - E_{m-1}; \ \overline{E}_{m-1} := E_{m-1} - E_m;$
$\qquad$ IF $(\overline{E}_{m-1} \subseteq g(E_L)) \wedge ((g(E_L) - \overline{E}_{m-1}) \subseteq E_{m-1}) \wedge (\overline{E}_m \subseteq g(E_R))$
$\qquad \wedge ((g(E_R) - \overline{E}_m) \subseteq E_m)$
$\qquad$ THEN $derivationExpr := derivationExpr \cup (E_m, (L, K, R), g);$
$\qquad$ ELSE RETURN $(derivationExpr, \ E_0);$
$\quad$ END;

Consider a sequence $G_0, \ldots, G_k$ of graphs and a rule set P as defined in Sec. 2. For every graph G_m (with $1 \le m \le k$), every rule $r = (L, K, R) \in P$, and every $g \in \mathcal{M}(r, n)$ it is checked by comparing the edges of G_{m-1} and G_m if r is applied w.r.t. g in the mth derivation step. The IF-condition assures that (1) the deleted edges of the mth derivation step are in $g(E_L)$; (2) the edges of $g(E_L)$ which are not deleted are in G_{m-1}; (3) the edges which are added in the mth derivation step are in $g(E_R)$; and (4) the edges of $g(E_R)$ which are not added are in G_m. For some rule sets the algorithm is ambiguous, but for the graph transformation unit $jobshop(M, J, pt, l)$ we receive a unique result.

By extracting a derivation expression from this result it is now possible to use GrGen.NET for the visualization of a solution. Fig. 7 shows an excerpt of a derivation with the 3×3 job-shop graph in Fig. 1 of Sec. 3 as initial graph consisting of the first and last two steps of 48 steps altogether. Obviously, the GrGen-output is much more transparent than the MiniSat-output in Fig. 6.

The above-mentioned example, with nine nodes total, yields a propositional formula with 9720861 literals and a cnf formula with 49113 variables and 9723633 clauses. This is already an optimized version of the formula generation algorithm which takes advantage of node typing to strongly restrict the number of possible matches. The solving time in MiniSat is 81.26 seconds. The formula turns out to be satisfiable for makespan 3.

5.2 Discussion

As illustrated above, SATaGraT searches for successful derivations via MiniSat and brings the corresponding output into a readable form by translating it into a derivation executed by GrGen.NET. Hence, in case of NP-hard problems SATaGraT can be used for proving and illustrating the existence of successful derivations. The generated propositional formulas are already pretty large even for small graphs and the conversion to cnf represents a bottleneck, both for time and for memory reasons. All of the four tested cnf converters (Funsat [7], Limboole [14], Logic2CNF [15], and Sugar [19]) take between 1.5 and 4+ hours to

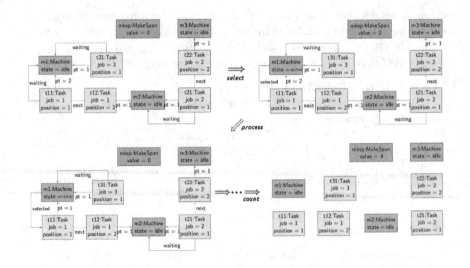

Fig. 7. An excerpt of a derivation

handle the cnf conversion for the above nine-node example, with Limboole being by far the fastest and Funsat the slowest. For the larger examples we tried, the cnf conversion was prematurely aborted due to insufficient memory. In order to handle not only small input graphs, optimization concerning both the size of the generated propositional formulas and the efficiency of the cnf conversion are needed. The propositional formula encodes all possible derivations of the given transformation unit. Hence, one idea to shrink the formulas is to drop non-successful derivations in advance. We are hopeful that analyzing the structure of the initial graph and the rules of the graph transformation unit will prove useful in achieving this kind of optimization. We have already implemented a matching optimization mentioned above that has reduced the original size of the formulas by up to 95 percent.

6   Conclusion

In this paper, we have proposed a combination of a SAT solver like MiniSat and a graph transformation engine like GrGen.NET. The hope is that this tool increases the chance to find solutions for instances of NP-hard problems that are modeled by graph transformation systems. It is known that SAT solvers are employed in chip design and verification with some success, and our idea has been to take up these positive experiences and to carry them over to graph transformation. We have designed and implemented the prototypical tool SATaGraT (*SAT solver assists graph transformation engine*) and have tested it with graph transformational models of various problems including the job-shop problem.

Our tentative approach meets three goals.

1. The translation of a special type of graph transformation units into propositional formulas as proposed in [13] is supplemented by a follow-up translation of variable assignments satisfying the formulas into successful derivations of the original graph transformation units.
2. Both translations are implemented prototypically by the SATaGraT system.
3. As the job-shop scheduling example demonstrates, a SAT solver may help to find solutions of NP-hard graph transformational problems. If the SAT solver within SATaGraT yields a result, it is exact. If the formula is unsatisfiable, there is no successful derivation. This implies that based on SATaGraT and a SAT solver, it is automatically decidable for the class of considered transformation units whether their semantics is empty. If it is not empty, a successful derivation is provided and can be executed and visualized on GrGen.NET. In contrast, if one searches for successful derivations directly on GrGen.NET, proper results are pure chance with very small probability in general.

In the future, the presented work should be further developed with respect to the following aspects.

- In our very first experiments the sizes of the input graphs are still rather small. The problem is that our translation of graph transformation units into propositional formulas yields very large numbers of clauses although the formulas are of polynomial size. Hence, further optimizations are needed to obtain a SATaGraT behavior that is competitive with the benchmarks in the literature.
- One such optimization may be achieved by using structural analysis on the input graph and the graph transformation unit to reduce the nondeterminism in the derivation and thus the size of the formula.
- The performance of SATaGraT should be checked against further graph transformational descriptions of NP-hard graph algorithms.
- In particular, it should be investigated whether SATaGraT may help to improve upper bounds found so far by non-exact methods applied to concrete problem instances.
- The presented approach should be compared with other methods for finding exact solutions for NP-hard problems such as fixed parameter algorithms (cf., e.g., [16]).

References

1. Biere, A., Cimatti, A., Clarke, E., Fujita, M., Zhu, Y.: Symbolic model checking using SAT procedures instead of BDDs. In: Irwin, M.J. (ed.) Proc. of the 36th Annual ACM/IEEE Design Automation Conference, pp. 317–320. ACM (1999)
2. Biere, A., Heule, M., van Maaren, H., Walsh, T. (eds.): Handbook of Satisfiability, Frontiers in Artificial Intelligence and Applications, vol. 185. IOS Press (2009)
3. Blazewicz, J., Ecker, K.H., Pesch, E., Schmidt, G., Weglarz, J. (eds.): Handbook on Scheduling: From Theory to Applications. Springer (2007)
4. Brucker, P.: Scheduling Algorithms. Springer (2007)

5. Eén, N., Sörensson, N.: An Extensible SAT-solver. In: Giunchiglia, E., Tacchella, A. (eds.) SAT 2003. LNCS, vol. 2919, pp. 502–518. Springer, Heidelberg (2004)
6. Ehrig, H., Ehrig, K., Prange, U., Taentzer, G.: Fundamentals of Algebraic Graph Transformation. Springer (2006)
7. Funsat (2011), `http://hackage.haskell.org/package/funsat-0.6.0` (accessed August 26, 2011)
8. Ganai, M., Gupta, A.: SAT-Based Scalable Formal Verification Solutions. Series on Integrated Circuits and Systems. Springer (2007)
9. Garey, M.R., Johnson, D.S.: Computers and Intractability: A Guide to the Theory of NP-Completeness. W. H. Freeman (1979)
10. Geiß, R., Kroll, M.: GrGen.NET: A Fast, Expressive, and General Purpose Graph Rewrite Tool. In: Schürr, A., Nagl, M., Zündorf, A. (eds.) AGTIVE 2007. LNCS, vol. 5088, pp. 568–569. Springer, Heidelberg (2008)
11. Kreowski, H.J., Kuske, S.: Graph transformation units with interleaving semantics. Formal Aspects of Computing 11(6), 690–723 (1999)
12. Kreowski, H.-J., Kuske, S., Rozenberg, G.: Graph Transformation Units – An Overview. In: Degano, P., De Nicola, R., Meseguer, J. (eds.) Concurrency, Graphs and Models. LNCS, vol. 5065, pp. 57–75. Springer, Heidelberg (2008)
13. Kreowski, H.-J., Kuske, S., Wille, R.: Graph Transformation Units Guided by a SAT Solver. In: Ehrig, H., Rensink, A., Rozenberg, G., Schürr, A. (eds.) ICGT 2010. LNCS, vol. 6372, pp. 27–42. Springer, Heidelberg (2010)
14. Limboole (2011), `http://fmv.jku.at/limboole/` (accessed August 26, 2011)
15. Logic2CNF (2011), `http://projects.cs.kent.ac.uk/projects/logic2cnf/trac/` (accessed August 26, 2011)
16. Niedermeier, R.: Invitation to Fixed-Parameter Algorithms. Oxford Lecture Series in Mathematics and its Applications, vol. 31. Oxford University Press, USA (2006)
17. Pinedo, M.: Scheduling: Theory, Algorithms, and Systems. Springer (2008)
18. Rozenberg, G. (ed.): Handbook of Graph Grammars and Computing by Graph Transformation, vol. 1: Foundations. World Scientific (1997)
19. Sugar (2011), `http://bach.istc.kobe-u.ac.jp/sugar/` (accessed August 26, 2011)
20. Tseitin, G.: On the complexity of derivation in propositional calculus. In: Studies in Constructive Mathematics and Mathematical Logic, Part 2, pp. 115–125 (1968); (reprinted in: Siekmann, J., Wrightson, G. (eds.): Automation of Reasoning, vol. 2, pp. 466–483. Springer (1983))

Locality in Reasoning
about Graph Transformations

Martin Strecker*

IRIT (Institut de Recherche en Informatique de Toulouse)
Université de Toulouse, France
http://www.irit.fr/~Martin.Strecker/

Abstract. This paper explores how to reason locally about global properties of graph transformations, in particular properties involving transitive closure of the edge relations of the graph. We show under which conditions we can soundly reduce reasoning about all nodes in the graph to reasoning about a finite set of nodes. We then give an effective procedure to turn this reduced problem into a Boolean satisfiability problem.

Keywords: Graph transformations, formal methods, verification.

1 Introduction

Proving the correctness of graph transformations is a notoriously hard problem. As opposed to transformations of tree-like structures, such as term rewriting or functional programming over inductively defined datatypes, there are no well-defined traversal strategies of a graph, and, still worse, multiple matchings of a graph rewriting rule are possible for a single position in a graph. This lack of structure leads to a combinatorial explosion which calls for tool support when proving properties of graph transformations.

Two major approaches have been developed: The *model checking* approach (embodied by [12,3,8,11,19]) considers the evolution of a given start graph under application of transformation rules. Typical questions of interest are whether certain invariants are maintained throughout all evolutions of a graph, or whether certain states are reachable. A downside of this approach is that, in principle, it only allows to talk about the effect of transformations on individual graphs and not about the correctness of rules in general. (However, in particular cases, the rules themselves may generate all interesting instances, such as in the red-black trees of [2].)

The *theorem proving* approach (which we follow in this paper) hoists pre-/postcondition calculi known from imperative programming to programs about graphs. Even before being able to reason about more complex graph transformation programs, one has to be able to prove properties of a single transformation step, *i. e.*, the consequences of the application of a single rule.

---

* Part of this research has been supported by the *Climt* project (ANR-11-BS02-016).

A. Schürr, D. Varró, and G. Varró (Eds.): AGTIVE 2011, LNCS 7233, pp. 169–181, 2012.

Thus, in [6], a relational approach to transformation verification is described and a coding in Event-B presented [17]. In [15], a natural deduction-like proof system for reasoning about graph transformations is given, [9] establishes a correspondence between categorical notions and satisfiability problems, however without providing automated verification procedures.

So far, in all this work, one question has remained unanswered: what is the global impact of applying a rule locally? Under which conditions can reasoning about a graph with an unbounded (possibly even infinite) number of nodes be reduced to reasoning about the finite number of nodes the rule is applied to? The global properties we are particularly interested in are connectedness and reachability, which, in principle, require inductive proofs. We especially concentrate on preservation of transitive closure of the form $r^* \subseteq r'^*$, where r resp. r' are the arc relations of a graph before resp. after the transformation. Finding good answers is essential for automating the verification of graph transformations. We will show when and how we can reduce the containment problem $r^* \subseteq r'^*$ to a problem of the form $r \subseteq r'$. This avoids reasoning about possibly unbounded paths and instead only considers a finite set of nodes and arcs that can be derived directly from the transformation rules. We thus arrive at a problem that can be decided by Boolean satisfiability checking.

We will further highlight the problem and give an informal overview of our approach in Section 2. We then describe our language for specifying graph transformations and the property language in Section 3. In Section 4, we show under which conditions we can separate a graph into an arbitrarily large "exterior" and a finite "interior", and in Section 5, we reduce the verification of a given property on this interior to a Boolean satisfiability problem. In Section 6, we conclude with an outlook on future work.

2   Problem Statement

Consider the example in Figure 1, describing a graph transformation rule (in the upper part) that deletes the edge (n_1, n_3) and instead inserts two new edges (n_1, n_2) and (n_2, n_3). When this transformation is embedded into a larger graph (in the lower part; the image of the rule in the graph has a darker shading), one can assert that if it is possible to go from a node x to a node y in the original graph, then this is also possible after transformation. More formally: if r is the edge relation and r^* is its reflexive-transitive closure, then $(x, y) \in r^*$ in the original graph implies $(x, y) \in r^*$ in the transformed graph, for arbitrary nodes x and y.

It might appear that this fact can be established by simply looking at the finite number of nodes of the transformation rule, without taking arbitrary x and y in the graph. Thus, the path $(n_1, n_3) \in r^*$ of the LHS of the rule can be constructed by the composition of (n_1, n_2) and (n_2, n_3) in the RHS, which seems to carry over to embeddings of this rule in larger graphs.

Such a reasoning is fallacious, as illustrated in Figure 2, which is taken from a case study describing the data flow in a communication network (a similar

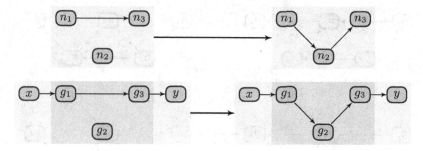

Fig. 1. Simple rerouting

example appears in [1]). The network consists of composite elements (bold borders in the figure) and simple elements. There are two kinds of relations: the flow relation (simple arrows) and the containment relation between simple and composite elements (dashed arrows). The reader should not be confused by the terminology: also a node representing a composite element is a regular node and not a subgraph.

The transformation rule describes the situation where diagrams with composite elements are flattened, by rerouting data flows to and from composite elements to their containing blocs (of course, the containment relation cannot be part of data paths). One can verify that data paths between non-composite elements are preserved within the rule: the path from n_1 to n_4 in the LHS also exists in the RHS. Unfortunately, data path preservation does not hold any more in a larger context, when considering nodes outside the image of the transformation rule: For example, the data path from g_1 to g_6 is lost by the transformation.

To make the statement ("paths between non-composite blocs") more formal, let us use three kinds of relations: r for edges between simple blocs (straight, non-dashed arrows), and r_{sc} (resp. r_{cs}) for edges from simple to composite (resp. composite to simple) blocs. Let us write $\circ$ for relation composition. The path preservation property now reads: if $(x, y) \in (r \cup (r_{cs} \circ r_{sc}))^*$ in the original graph for arbitrary x, y, then also in the transformed graph.

Why does local reasoning about a rule carry over to the entire graph in the first, but not in the second case? One of the causes appears to be relation composition, which hides an existential quantifier: $(x, y) \in (r_{cs} \circ r_{sc})$ means $\exists z.(x, z) \in r_{cs} \wedge (z, y) \in r_{sc}$, and there is no way to tell whether this z is one of the nodes inside the image of the transformation rule, or whether it lies outside of it. We will therefore omit relation composition from the relational language of Section 5.1, but definitely show that local reasoning is safe for the fragment defined below.

3   Representing and Reasoning about Transformations

Our way of representing and reasoning about graph transformations is described in more detail in [18], and we therefore only summarize the elements that are

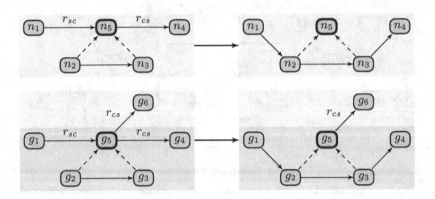

Fig. 2. Complex rerouting

most important for an understanding of this article. This section might be
skipped on a first reading and only be consulted for the terminology used later
on.

The approach is entirely logical and consists of coding all relevant notions of
graph transformations in a higher-order proof assistant, such as rules, matching,
morphisms and rewriting itself. (In our case, the Isabelle [14] system is used, and
some of the definitions below are directly extracted from the Isabelle sources.)
This gives a fine control over the transformation process, such as the properties
of the morphisms mapping rules into graphs. For example, we assume throughout
this article that these morphisms are injective, but this property can easily be
configured differently, however with a non-negligible impact on efficiency of the
reasoning procedures.

A graph is defined to be a record consisting of a set of nodes and a set of
edges (pairs of nodes), indexed by an edge type $'et$. Furthermore, nodes can be
given a node type:

record $('nt, 'et, 'h)$ *graph* =
  nodes :: *obj set*
  edges :: $'et \Rightarrow (obj * obj)$ *set*
  nodetp :: $obj \Rightarrow 'nt$

A graph transformation consists of an applicability condition that defines a
matching in a target graph, and an effect that specifies how the matched nodes
and edges are transformed: which nodes and edges have to be deleted and which
have to be generated, and how the newly generated nodes have to be typed. In
order to be able to manipulate the applicability condition syntactically, we do
not express it with the aid of the built-in logic of our proof assistant, but rather
use a special-purpose syntax that we do not spell out any further here. As an
illustration, here is the definition that specifies the transformation of Figure 1.
The precondition *rerouting-precond* expresses that there is a type of nodes, *Node*,
and a relation r, that the nodes $n1$, $n2$, $n3$ have type *Node* and there is an edge
of type r between $n1$ and $n3$. The transformation itself specifies which edges are

deleted (*e-del*) and generated (*e-gen*). This specification is indexed by the node relation, hence the lambda-abstraction over *et*. No nodes are deleted or added in this example.

datatype *nodetp = Node*
datatype *edgetp = r*

definition
  rerouting-precond :: *nat* ⇒ *nat* ⇒ *nat* ⇒ (*nodetp, edgetp,'h*) *path-form* **where**
  rerouting-precond n1 n2 n3 = (*Conjs-form* [
    (*S-form* (*Type-set Node*) *n1*), (*S-form* (*Type-set Node*) *n2*),
    (*S-form* (*Type-set Node*) *n3*), (*P-form* (*Edge-pth r*) *n1 n3*)])

definition *rerouting* :: (*nodetp, edgetp, 'h*) *graphtrans* **where**
  rerouting =
  (| *appcond* = *rerouting-precond 1 2 3*,
    n-del = {},   *n-gen* = {},
    e-del = λ *et*. {(1,3)}, *e-gen* = λ *et*. {(1,2), (2,3)},
    n-gentp = *empty*)

A morphism is a map from nodes (in the transformation rule) to nodes (in the target graph):

types *graphmorph = nat* ⇒ *obj option*

We now have all ingredients for defining the function *apply-graphtrans-rel* that applies a graph transformation under a given morphism to a graph and produces a transformed graph (see [18] for more details):

consts *apply-graphtrans-rel* ::
  [('*nt, 'et, 'h*) *graphtrans, graphmorph,* ('*nt, 'et, 'h*) *graph,* ('*nt, 'et, 'h*) *graph*]
  ⇒ *bool*

Usually, we are not interested in the behavior of a transformation for one particular morphism, but for any morphism that satisfies the applicability condition. The following relation abstracts away from the morphism and just describes that a graph *gr* is transformed into a graph *gr'* by a graph transformation *gt*:

definition *apply-transfo-rel* ::
  [('*nt, 'et, 'h*) *graphtrans,* ('*nt, 'et, 'h*) *graph,* ('*nt, 'et, 'h*) *graph*] ⇒ *bool* **where**
  apply-transfo-rel gt gr gr' = *apply-graphtrans-rel gt* (*select-morph gt gr*) *gr gr'*

The proof obligations we set out to prove typically have the form: *if the transformation transforms gr into gr', then P (gr, gr') holds*, where *P* is a predicate putting into correspondence the node and edge sets of *gr* and *gr'*. For example, the path preservation property of Figure 1 is stated as:

(*applicable-transfo rerouting gr* ∧ *apply-transfo-rel rerouting gr gr'*)
  ⟹ (*edges gr r*)* ⊆ (*edges gr' r*)*
More specifically, the properties *P* we will examine in the following are simple preservation properties *r* ⊆ *s* or path preservation properties *r** ⊆ *s**, where *r* represents the edge relation of *gr* and *s* the edge relation of *gr'* ("nodes connected

in gr are also connected in gr'''), or inversely ("nodes connected in gr' were already connected in gr'', rather expressing a preservation of separation).

The way the transformations are defined, references to gr' can be eliminated. To illustrate this point, let us continue with our example. Unfolding definitions and carrying out some simplifications, our proof obligation reduces to the following (hypotheses are contained in ⟦ ... ⟧, and some inessential ones have been removed):

⟦$n1 \in nodes\ gr$; $n2 \in nodes\ gr$; $n3 \in nodes\ gr$; $nodetp\ gr\ n1 = Node$;
 $nodetp\ gr\ n2 = Node$; $nodetp\ gr\ n3 = Node$; $(n1, n3) \in edges\ gr\ r$⟧
$\Longrightarrow (edges\ gr\ r)^*$
$\quad\quad \subseteq (insert\ (n1, n2)\ (insert\ (n2, n3)\ (edges\ gr\ r - \{(n1, n3)\})))^*$

4   Graph Decompositions

The essential step of our approach consists in splitting up a graph into an interior region (the image of nodes of a rule in a graph under a given graph morphism), and an exterior. Since we are here mostly concerned with the node relations of a graph, we define the interior and exterior of a relation r with respect to a region A as follows:

definition *interior-rel A r = r ∩ (A × A)*
definition *exterior-rel A r = r − (A × A)*

Visually speaking, the interior of a relation wrt. a region A comprises all arcs connecting nodes a_1, a_2 both of which are contained in A, whereas the exterior consists of the arcs connected to at least one node e outside of A.

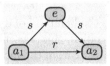

Fig. 3. Interior (dark shade) and exterior (light shade) of a relation

The interior and exterior of a relation add up to the whole relation again:

lemma *interior-union-exterior*: *interior-rel A r ∪ exterior-rel A r = r*

and from this, we obtain by simple set-theoretic reasoning a first decomposition lemma:

lemma *interior-exterior-decompos-subset-equal*:
  (*exterior-rel A r ⊆ exterior-rel A s ∧ interior-rel A r ⊆ interior-rel A s*)
  = (*r ⊆ s*)

When read from right to left, this lemma reduces reasoning of a relation containment property of the form $r \subseteq s$ to reasoning about the interior and exterior of these relations.

Our principal interest in this paper is reasoning about preservation of (reflexive) transitivity properties, *i. e.,* properties of the form $r^* \subseteq s^*$ or $r^+ \subseteq s^+$. Even though applicable in principle, the above decomposition lemma is of no much use, as will become apparent in Section 5. We therefore provide the following specialized form (a similar lemma holds for the transitive closure as well):

lemma *interior-exterior-decompos-rtrancl-subset*:
$(exterior\text{-}rel\ A\ r)^* \subseteq (exterior\text{-}rel\ A\ s)^* \wedge (interior\text{-}rel\ A\ r)^* \subseteq (interior\text{-}rel\ A\ s)^*$
$\implies r^* \subseteq s^*$

Its correctness can be seen by using the property $(I^* \cup E^*)^* = (I \cup E)^*$ of transitive closures and lemma *interior-union-exterior* to show that $((interior\text{-}rel\ A\ r)^* \cup (exterior\text{-}rel\ A\ r)^*)^* = r^*$ and then use simple set-theoretic reasoning.

As witnessed by this lemma, proving $r^* \subseteq s^*$ by decomposition into interior and exterior is sound, but contrary to *interior-exterior-decompos-subset-equal,* the above lemma is an implication and not an equality. One may therefore wonder whether this decomposition does not make us lose completeness, by possibly reducing a provable to an unprovable statement.

Indeed, the inverse does not hold in general. For example, $r^* \subseteq s^* \implies$ $(interior\text{-}rel\ A\ r)^* \subseteq (interior\text{-}rel\ A\ s)^*$ is not satisfied for $r = \{(a_1, a_2)\}$, $s = \{(a_1, e), (e, a_2)\}$ and $A = \{a_1, a_2\}$. Note that in this case, *interior-rel A* $s = \{\}$, whereas *interior-rel A* $r = \{(a_1, a_2)\}$ (also see Figure 3).

However, by choosing the interior large enough, we obtain the inverse of lemma *interior-exterior-decompos-rtrancl-subset* (here, *Field* is the union of the domain and the range of a relation):

lemma *decompos-complete-interior*:
$Field\ s \subseteq A \implies r^* \subseteq s^* \implies (interior\text{-}rel\ A\ r)^* \subseteq (interior\text{-}rel\ A\ s)^*$
lemma *decompos-complete-exterior*:
$Field\ r \subseteq A \implies r^* \subseteq s^* \implies (exterior\text{-}rel\ A\ r)^* \subseteq (exterior\text{-}rel\ A\ s)^*$

In practice, this means that the interior has to comprise all known elements of the relations r and s (differently said: all free variables occurring in them).

We have thus recovered completeness of our decomposition for a class of graphs that comprises also the graphs for which this decomposition is in fact trivial (*i. e., Field* $s \subseteq A$ and *Field* $r \subseteq A$ hold). It also highlights the problem of performing this decomposition if the fields of the involved relations are not entirely known, as in the case of relation composition.

Let us conclude this section by applying the above procedure to our example. Since the variables *n1*, *n2*, *n3* are free in our goal, we choose $A = \{n1, n2, n3\}$ and obtain two new subgoals, the first having the conclusion

$(exterior\text{-}rel\ \{n1,\ n2,\ n3\}\ (edges\ gr\ r))^*$
$\subseteq (exterior\text{-}rel\ \{n1,\ n2,\ n3\}$
$\quad (insert\ (n1,\ n2)\ (insert\ (n2,\ n3)\ (edges\ gr\ r - \{(n1,\ n3)\}))))^*$

and the second one being similar, for *interior-rel.*

5   Reduction to Boolean Satisfiability

As indicated before, we are interested in proving properties of the form $r \subseteq s$ or $r^* \subseteq s^*$, and we have seen in Section 2 that some relation constructors, such as relation composition, are problematic. In Section 5.1, we therefore give a more precise characterisation of the structure of the relation expressions r and s we will use in the following.

After decomposition of the graph into an exterior and an interior, we are left to simplifying each of these parts separately. In Section 5.2 and Section 5.3 respectively, we will see that for our relation algebra, the resulting problems are indeed decidable.

5.1   Relation Expressions

We assume that our relational expressions r are built up inductively according to the following grammar:

$$
\begin{aligned}
r ::= &\; r_b \\
| &\; \{\} \\
| &\; insert \; (n_1, n_2) \; r \\
| &\; r \cup r \\
| &\; r \cap r \\
| &\; r - r
\end{aligned}
$$

Here, r_b are basic, non-interpreted relations, and n_1, n_2 are variables representing node names. Our grammar contains set difference $A - B$ and not simply set complement $- B$ because the former behaves better with some of the reductions in Section 5.2 and Section 5.3.

We assume that $FV(r)$ returns all the free variables in r, $i.\,e.,$ the set of all the n_1 and n_2 occurring in $insert$-expressions of r.

It should be noted that this grammar enables to capture at least the effects of transformations that arise from application of rules as defined in Section 3.

5.2   Reduction of the Exterior

Decomposition of Sec. 4 has left us with a first subgoal of the form $(exterior\text{-}rel \; A \; r) \subseteq (exterior\text{-}rel \; A \; s)$ or $(exterior\text{-}rel \; A \; r)^* \subseteq (exterior\text{-}rel \; A \; s)^*$. We now exhaustively apply the following rewrite rules (technically, in our framework, they are added to Isabelle's simplification set):

$exterior\text{-}rel \; A \; \{\} = \{\}$
$exterior\text{-}rel \; A \; (insert \; (x, \; y) \; r) = (\{(x, \; y)\} - (A \times A)) \cup \; exterior\text{-}rel \; A \; r$
$exterior\text{-}rel \; A \; (r \cap s) = (exterior\text{-}rel \; A \; r) \cap (exterior\text{-}rel \; A \; s)$
$exterior\text{-}rel \; A \; (r \cup s) = (exterior\text{-}rel \; A \; r) \cup (exterior\text{-}rel \; A \; s)$
$exterior\text{-}rel \; A \; (r - s) = (exterior\text{-}rel \; A \; r) - (exterior\text{-}rel \; A \; s)$

and furthermore the simplification $\{(x,\ y)\} - (A \times A) = \{\}$ if $x,y \in A$, as well as trivial simplifications of operations with the empty set.

An easy inductive argument shows that, if $FV(r) \subseteq A$, these simplifications reduce any expression *exterior-rel A r* to a Boolean combination, consisting of operators $\cup$, $\cap$, $-$, and only involving expressions *exterior-rel A r_b*, where the r_b are basic relations.

If our original goal had the form $(exterior\text{-}rel\ A\ r) \subseteq (exterior\text{-}rel\ A\ s)$, we are now left with a goal $R \subseteq S$, with R and S such Boolean combinations. Since the basic r_b are uninterpreted, this goal can be proved or disproved with propositional reasoning.

If our original goal had the form $(exterior\text{-}rel\ A\ r)^+ \subseteq (exterior\text{-}rel\ A\ s)^+$, we are now left with a goal $R^+ \subseteq S^+$, which we can reduce to $R \subseteq S$, using the fact that (reflexive) transitive closure is monotonic, and then proceed as before. This reduction is obviously sound. It is also complete: if $R \subseteq S$ is not provable, any counterexample can be turned into a counterexample in which the basic expressions *exterior-rel A r_b* are interpreted as either empty or non-empty over a one-element domain. Over this domain, a relation and its transitive closure coincide, so that we obtain a counterexample of $R^+ \subseteq S^+$. A slightly more involved argument also holds for reflexive-transitive closure.

5.3   Reduction of the Interior

We now show how we can verify the properties in the interior, for goals having the form $(interior\text{-}rel\ A\ r) \subseteq (interior\text{-}rel\ A\ s)$ or $(interior\text{-}rel\ A\ r)^* \subseteq (interior\text{-}rel\ A\ s)^*$. Even though it appears intuitively plausible that these questions are decidable for a finite interior A, it is not sufficient to use traditional graph algorithms, because our relations r and s are composite.

For simplification, we proceed in several stages:

1. we push inside applications of *interior-rel*, leaving applications of the form *interior-rel A r_b* for basic relations r_b;
2. we explicitly calculate *interior-rel A r_b*, which completely eliminates all occurrences of *interior-rel*;
3. if necessary, we explicitly calculate the (reflexive) transitive closure of the sets thus obtained, which essentially gives a propositional problem;
4. we solve the problem with a propositional solver. ·

For the first stage, we use the following rewrite rules:

interior-rel A $\{\} = \{\}$
interior-rel A $(insert\ (x,\ y)\ r) = (\{(x,\ y)\} \cap (A \times A)) \cup (interior\text{-}rel\ A\ r)$
interior-rel A $(r \cap s) = (interior\text{-}rel\ A\ r) \cap (interior\text{-}rel\ A\ s)$
interior-rel A $(r \cup s) = (interior\text{-}rel\ A\ r) \cup (interior\text{-}rel\ A\ s)$
interior-rel A $(r - s) = (interior\text{-}rel\ A\ r) - (interior\text{-}rel\ A\ s)$

Note that A is a finite, concrete set, so that we can further simplify with:

$x \in A \land y \in A \Longrightarrow (\{(x, y)\} \cap (A \times A)) = \{(x, y)\}$
$x \notin A \lor y \notin A \Longrightarrow (\{(x, y)\} \cap (A \times A)) = \{\}$

At the end of this first step, we are now left with a Boolean combination of relations of the style *interior-rel A* r_b, where r_b is basic. The following equality:

$r \cap (insert\ a\ A) \times (insert\ a\ A) =$
$(r \cap (A \times A)) \cup (r \cap (\{a\} \times (insert\ a\ A))) \cup (r \cap ((insert\ a\ A) \times \{a\}))$

is the basis for a simplification scheme that recurses over A and eliminates all *interior-rel*:

> *interior-rel* $\{\}$ $r = \{\}$
> *interior-rel* $(insert\ a\ A)\ r =$
> $(interior\text{-}rel\ A\ r) \cup$
> $(interior\text{-}rel\text{-}elem\text{-}r\ a\ (insert\ a\ A)\ r) \cup (interior\text{-}rel\text{-}elem\text{-}l\ a\ (insert\ a\ A)\ r)$

Here, we have defined

definition *interior-rel-elem-r* $a\ B\ r = r \cap (\{a\} \times B)$
definition *interior-rel-elem-l* $b\ A\ r = r \cap (A \times \{b\})$

with the following reductions:

> *interior-rel-elem-r* $a\ \{\}\ r = \{\}$
> *interior-rel-elem-r* $a\ (insert\ b\ B)\ r =$
> $(if\ (a, b) \in r\ then\ \{(a, b)\}\ else\ \{\}) \cup interior\text{-}rel\text{-}elem\text{-}r\ a\ B\ r$
> *interior-rel-elem-l* $b\ \{\}\ r = \{\}$
> *interior-rel-elem-l* $b\ (insert\ a\ A)\ r =$
> $(if\ (a, b) \in r\ then\ \{(a, b)\}\ else\ \{\}) \cup interior\text{-}rel\text{-}elem\text{-}l\ b\ A\ r$

Before taking the third step, we have to massage the goal by if-splitting, thus distinguishing, in the above rules, between the cases $(a, b) \in r$ and $(a, b) \notin r$. This is a source of a considerable combinatorial explosion: for each basic relation r_b occurring in the goal, and for n nodes, we potentially get $2^{(n^2)}$ combinations, the problem being aggravated by the fact that the relations are not directed. We will come back to this point in Section 6.

To complete the transformation, we symbolically compute the closure. We show the recursive equation for transitive closure:

$(v, w) \in (insert\ (y, x)\ r)^+ =$
$((v, w) \in r^+ \lor ((v = y) \lor (v, y) \in r^+) \land ((x = w) \lor (x, w) \in r^+))$

After these transformations, we obtain a Boolean combination of

- membership in an elementary relation: $(x, y) \in r$ (or their negation)
- (in)equalities $x = y$ or $x \neq y$ between nodes

This fragment can readily be decided by standard propositional solvers.

6   Conclusions

We have presented a method of automatically proving properties of graph transformations in a restricted relational language with transitive closure, by decomposing the abstract graph in which a rule is to be applied in an interior and an exterior region on which the proof problems can be verified by essentially propositional means. Seen from a different angle: the present paper establishes a general framework in which the difficultly automizable inductive proofs arising from transitive closure operations have already been carried out on the meta-level. For a particular application, it therefore suffices to use finitary proof methods.

The transformations in Section 5.2 might give rise to problems of set containment for Boolean set operations, which is in principle NP-complete, but seems to be very efficient in practice. The reason is that the simplifications in Section 5.2 only verify that a decomposition has been applied correctly.

However, in their current form, the transformations in Section 5.3 are of exponential complexity. Intuitively, this is because we have to check for the presence of any possible combination of edges in the graph in which the rule has to be applied. It seems difficult to see how to do better, because contrary to tree-like structures, graph transformations allow for a large number of matchings. We are currently working on a proof of the complexity of this decision problem.

We can extend our work in different directions: On the practical side, we would like to develop simplification strategies that perform essential reductions before case splitting introduces a combinatorial explosion.

On the theoretical side, it is interesting to further explore the boundary between decidable and undecidable fragments, hopefully leading to a more expressive relational fragment than the one of Section 5.1. Several logics for reasoning about graphs [5] and for pointer structures in programs have been proposed, and an in-depth comparison still has to be done. The decidable fragment in [10], based on transitive closure logic, only allows for a single edge relation, whereas the logical language in [20] assumes a set of distinguished starting points for talking about reachability, but does not examine reachability between arbitrary points.

The decidable logic described in [13] concentrates on the preservation of data structure variants (such as doubly-linked lists) in pointer manipulating programs, but does not allow to reason about global graph properties.

Separation Logic (SL) [16] is a specialized logic for reasoning in a pre-/ postcondition style about pointer-manipulating programs. It advocates splitting the heap into disjoint areas, thus reducing the complexity of reasoning about pointer manipulations which may have a global impact due to aliasing. Just like Hoare logic, SL as such does not aim at being a decidable logic. The decidable fragment described in [4] is so weak that it is not closed under the heap-manipulating operations of a programming language. It only enables to describe simple properties of lists, but not more general properties of connectivity.

The relation of SL and graph rewriting is discussed in [7], however without reference to particular decision procedures.

Probably the best way to come to terms with the complexity of graph rewriting is to identify particular classes of graphs that are better behaved. This might be tree-like graphs or at least graphs in which the non-determinism of rule application is strongly restricted.

Acknowledgements. I am grateful for my colleagues Ralph Matthes, Christian Percebois and Hanh-Ni Tran for discussions about the problem of locality in graph rewriting and for helpful feedback on preliminary versions of this article.

References

1. Asztalos, M., Lengyel, L., Levendovszky, T.: Towards automated, formal verification of model transformations. In: Proc. of the 3rd International Conference on Software Testing, Verification, and Validation, pp. 15–24. IEEE Computer Society, Los Alamitos (2010)
2. Baldan, P., Corradini, A., Esparza, J., Heindel, T., König, B., Kozioura, V.: Verifying red-black trees. In: Proc. of Concurrent Systems with Dynamic Allocated Heaps, COSMICAH 2005 (2005); Proceedings available as report RR-05-04 (Queen Mary, University of London)
3. Baldan, P., Corradini, A., König, B.: A framework for the verification of infinite-state graph transformation systems. Information and Computation 206(7), 869–907 (2008)
4. Berdine, J., Calcagno, C., O'Hearn, P.: A decidable fragment of separation logic. In: Lodaya, K., Mahajan, M. (eds.) FSTTCS 2004. LNCS, vol. 3328, pp. 97–109. Springer, Heidelberg (2004)
5. Caferra, R., Echahed, R., Peltier, N.: A term-graph clausal logic: Completeness and incompleteness results. Journal of Applied Non-classical Logics 18(4), 373–411 (2008)
6. da Costa, S.A., Ribeiro, L.: Formal verification of graph grammars using mathematical induction. In: Proc. of the 11th Brazilian Symposium on Formal Methods (SBMF 2008). ENTCS, vol. 240, pp. 43–60. Elsevier (2009)
7. Dodds, M.: Graph Transformation and Pointer Structures. Ph.D. thesis, University of York (2008)
8. Ghamarian, A.H., de Mol, M.J., Rensink, A., Zambon, E., Zimakova, M.V.: Modelling and analysis using GROOVE. Tech. Rep. TR-CTIT-10-18, Centre for Telematics and Information Technology, University of Twente, Enschede (2010)
9. Habel, A., Pennemann, K.H.: Correctness of high-level transformation systems relative to nested conditions. Mathematical Structures in Computer Science 19(02), 245–296 (2009)
10. Immerman, N., Rabinovich, A., Reps, T., Sagiv, M., Yorsh, G.: The Boundary Between Decidability and Undecidability for Transitive-Closure Logics. In: Marcinkowski, J., Tarlecki, A. (eds.) CSL 2004. LNCS, vol. 3210, pp. 160–174. Springer, Heidelberg (2004)
11. Kastenberg, H.: Graph-Based Software Specification and Verification. Ph.D. thesis, University of Twente (2008), http://eprints.eemcs.utwente.nl/13615/
12. König, B., Kozioura, V.: Augur 2 – A new version of a tool for the analysis of graph transformation systems. In: Proc. of the 5th Int. Workshop on Graph Transformation and Visual Modeling Techniques. ENTCS, vol. 211, pp. 201–210. Elsevier (2008)
13. McPeak, S., Necula, G.C.: Data Structure Specifications via Local Equality Axioms. In: Etessami, K., Rajamani, S.K. (eds.) CAV 2005. LNCS, vol. 3576, pp. 476–490. Springer, Heidelberg (2005)
14. Nipkow, T., Paulson, L.C., Wenzel, M.: Isabelle/HOL. LNCS, vol. 2283. Springer, Heidelberg (2002)

15. Pennemann, K.H.: Resolution-Like Theorem Proving for High-Level Conditions. In: Ehrig, H., Heckel, R., Rozenberg, G., Taentzer, G. (eds.) ICGT 2008. LNCS, vol. 5214, pp. 289–304. Springer, Heidelberg (2008)

16. Reynolds, J.C.: Separation logic: A logic for shared mutable data structures. In: Proc. of the 17th Annual IEEE Symposium on Logic in Computer Science, pp. 55–74. IEEE Computer Society (2002)

17. Ribeiro, L., Dotti, F.L., da Costa, S.A., Dillenburg, F.C.: Towards theorem proving graph grammars using Event-B. In: Ermel, C., Ehrig, H., Orejas, F., Taentzer, G. (eds.) Proc. of the International Colloquium on Graph and Model Transformation 2010. ECEASST, vol. 30 (2010)

18. Strecker, M.: Modeling and verifying graph transformations in proof assistants. In: Mackie, I., Plump, D. (eds.) International Workshop on Computing with Terms and Graphs. ENTCS, vol. 203, pp. 135–148. Elsevier (2008)

19. Varró, D., Balogh, A.: The model transformation language of the VIATRA2 framework. Science of Computer Programming 68(3), 214–234 (2007)

20. Yorsh, G., Rabinovich, A.M., Sagiv, M., Meyer, A., Bouajjani, A.: A logic of reachable patterns in linked data-structures. The Journal of Logic and Algebraic Programming 73(1-2), 111–142 (2007)

Contextual Hyperedge Replacement

Frank Drewes[1], Berthold Hoffmann[2], and Mark Minas[3]

[1] Umeå Universitet, Sweden
[2] DFKI Bremen and Universität Bremen, Germany
[3] Universität der Bundeswehr München, Germany

Abstract. In model-driven design, the structure of software is commonly specified by meta-models like UML class diagrams. In this paper we study how graph grammars can be used for this purpose, using state-charts as an example. We extend context-free hyperedge-replacement—which is not powerful enough for this application—so that rules may not only access the nodes attached to the variable on their left-hand side, but also nodes elsewhere in the graph. Although the resulting notion of contextual hyperedge replacement preserves many properties of the context-free case, it has considerably more generative power—enough to specify software models that cannot be specified by class diagrams.

1 Introduction

Graphs are ubiquitous in science and beyond. When graph-like diagrams are used to model system development, it is important to define precisely whether a diagram is a valid model or not. Often, models are defined as the valid instantiations of a *meta-model*, e.g., the valid object diagrams for a class diagram in UML. A meta-model is convenient for capturing requirements as it can be refined gradually. It is easy to check whether a given model is valid for a meta-model. However, it is not easy to construct valid sample models for a meta-model, and they give no clue how to define transformations on all valid models. Also, their abilities to express structural properties (like hierarchical nesting) are limited; constraints (e.g., in the logic language OCL) have to be used for more complicated properties like connectedness.

In contrast to meta-models, *graph grammars* derive sets of graphs constructively, by applying rules to a start graph. This kind of definition is strict, can easily produce sample graphs by derivation, and its rules provide for a recursive structure to define transformations on the derivable graphs. However, it must not be concealed that validating a given graph, by *parsing*, may be rather complex.

General graph grammars generate all recursively enumerable sets of graphs [16] so that there can be no parsing algorithm. Context-free graph grammars based on node replacement or hyperedge replacement [6] do not have the power to generate graphs of general connectivity, like the language of all graphs, of all acyclic, and all connected graphs etc. We conclude that practically useful kinds of graph grammars should lie in between context-free and general ones. We take hyperedge replacement as a solid basis for devising such grammars, as it has

A. Schürr, D. Varró, and G. Varró (Eds.): AGTIVE 2011, LNCS 7233, pp. 182–197, 2012.

a comprehensive theory, and is very simple: A step removes a variable (represented as a hyperedge) and glues the fixed ordered set of nodes attached to it to distinguished nodes of a graph. The authors have been working on several extensions of hyperedge replacement. *Adaptive star replacement* [2], devised with D. Janssens and N. Van Eetvelde, allows variables to be attached to arbitrary, unordered sets of nodes. Its generative power suffices to define sophisticated software models like program graphs [3]. Nevertheless, it inherits some of the strong properties of hyperedge replacement. Unfortunately, adaptive star rules tend to have many edges, which makes them hard to understand—and to construct. Therefore the authors have devised *contextual graph grammars*, where variables still have a fixed, ordered set of attached nodes, but replacement graphs may be glued, not only with these attachments, but also with nodes occurring elsewhere in the graph, which have been generated in earlier derivation steps [11]. As we shall show, their generative power suffices to define non-context-free models. Typically, contextual rules are only modest extensions of hyperedge replacement rules, and are significantly easier to write and understand than adaptive star rules. This qualifies contextual hyperedge grammars as a practical notation for defining software models. When we add application conditions to contextual rules, as we have done in [11], even subtler software models can be defined. Since conditions are a standard concept of graph transformation, which have been used in many graph transformation systems (see, e.g., PROGRES [15]), such rules are still intuitive.

This paper aims to lay a fundament to the study of contextual hyperedge replacement. So we just consider grammars without application conditions for the moment, as our major subjects of comparison, context-free hyperedge replacement and adaptive star replacement, also do not have them. With context-free hyperedge replacement, contextual hyperedge replacement shares decidability results, characterizations of their generated language, and the existence of a parsing algorithm. Nevertheless, it is powerful enough to make it practically useful for average structural models.

The remainder of this paper is structured as follows. In Section 2 we introduce contextual hyperedge replacement grammars and give some examples. In particular, we discuss a grammar for statecharts in Section 3. Normal forms for these grammars are presented in Section 4. In Section 5 we show some of their limitations w.r.t. language generation, and sketch parsing in Section 6. We conclude with some remarks on related and future work in Section 7.

2   Graphs, Rules, and Grammars

In this paper, we consider directed and labeled graphs. We only deal with abstract graphs in the sense that graphs that are equal up to renaming of nodes and edges are not distinguished. In fact, we use hypergraphs with a generalized notion of edges that may connect any number of nodes, not just two. Such edges will also be used to represent variables in graphs and graph grammars.

We consider labeling alphabets $C = \dot{C} \uplus \bar{C} \uplus X$ that are sets whose elements are the *labels* (or "*colors*") for nodes, edges, and variables, with an *arity* function $arity \colon \bar{C} \uplus X \to \dot{C}^*$.[1]

A *labeled hypergraph over* C (*graph*, for short) $G = \langle \dot{G}, \bar{G}, att_G, \ell_G, \bar{\ell}_G \rangle$ consists of disjoint finite sets $\dot{G}$ of *nodes* and $\bar{G}$ of *hyperedges* (*edges*, for short) respectively, a function $att_G \colon \bar{G} \to \dot{G}^*$ that *attaches* sequences of pairwise distinct nodes to edges so that $\ell_G^*(att_G(e)) = arity(\bar{\ell}_G(e))$ for every edge $e \in \bar{G}$,[2] and *labeling* functions $\ell_G \colon \dot{G} \to \dot{C}$ and $\bar{\ell}_G \colon \bar{G} \to \bar{C} \uplus X$. Edges are called *variables* if they carry a variable name as a label; the set of all graphs over C is denoted by $\mathcal{G}_C$.

For a graph G and hyperedge $e \in \bar{G}$, we denote by $G - e$ the graph obtained by removing e from G. Similarly, for $v \in \dot{G}$, $G - v$ is obtained by removing v from G (together with all edges attached to v).

Given graphs G and H, a *morphism* $m \colon G \to H$ is a pair $m = \langle \dot{m}, \bar{m} \rangle$ of functions $\dot{m} \colon \dot{G} \to \dot{H}$ and $\bar{m} \colon \bar{G} \to \bar{H}$ that preserves labels and attachments:

$$\ell_H \circ \dot{m} = \ell_G, \bar{\ell}_H \circ \bar{m} = \bar{\ell}_G, att_H(\bar{m}(e)) = \dot{m}^*(att_G(e)) \text{ for every } e \in \bar{G}$$

As usual, a morphism $m \colon G \to H$ is *injective* if both $\dot{m}$ and $\bar{m}$ are injective.

The replacement of variables in graphs by graphs is performed by applying a special form of standard double-pushout rules [5].

Definition 1 (Contextual Rule). A *contextual rule* (*rule*, for short) $r = (L, R)$ consists of graphs L and R over C such that

- the *left-hand side* L contains exactly one edge x, which is required to be a variable (i.e., $\bar{L} = \{x\}$ with $\bar{\ell}_L(x) \in X$) and
- the *right-hand side* R is an arbitrary supergraph of $L - x$.

Nodes in L that are not attached to x are the *contextual nodes* of L (and of r); r is *context-free* if it has no contextual nodes. (Context-free rules are known as hyperedge replacement rules in the literature [7].)

Let r be a contextual rule as above, and consider some graph G. An injective morphism $m \colon L \to G$ is called a *matching* for r in G. The *replacement* of the variable $m(x) \in G$ by R (via m) is the graph H obtained from the disjoint union of $G - m(x)$ and R by identifying every node $v \in \dot{L}$ with $m(v)$. We write this as $H = G[R/m]$.

Note that contextual rules are equivalent to contextual star rules as introduced in [11], however without application conditions.

The notion of rules introduced above gives rise to a class of graph grammars. We call these grammars contextual hyperedge-replacement grammars, or briefly contextual grammars.

---

[1] A^* denotes finite sequences over a set A; the empty sequence is denoted by ε.

[2] For a function $f \colon A \to B$, its extension $f^* \colon A^* \to B^*$ to sequences is defined by $f^*(a_1, \ldots, a_n) = f(a_1) \ldots f(a_n)$, for all $a_i \in A$, $1 \leqslant i \leqslant n$, $n \geqslant 0$.

Definition 2 (Contextual Hyperedge-Replacement Grammar). A *contextual hyperedge-replacement grammar* (*contextual grammar*, for short) is a triple $\Gamma = \langle C, R, Z \rangle$ consisting of a finite labeling alphabet C, a finite set R of rules, and a start graph $Z \in \mathcal{G}_C$.

If R contains only context-free rules, then Γ is a *hyperedge replacement grammar*. We let $G \Rightarrow_R H$ if $H = G[R/m]$ for some rule (L, R) and for a matching $m: L \to G$. Now, the language generated by Γ is given by

$$\mathcal{L}(\Gamma) = \{G \in \mathcal{G}_{C \backslash X} \mid Z \Rightarrow_R^* G\}.$$

Contextual grammars Γ and Γ' are *equivalent* if $\mathcal{L}(\Gamma) = \mathcal{L}(\Gamma')$. The classes of graph languages generated by hyperedge-replacement grammars and contextual grammars are denoted by HR and CHR, respectively.

Notation (Drawing Conventions for Graphs and Rules). Graphs are drawn as in Figure 2 and Figure 4. Circles and boxes represent nodes and edges, respectively. The text inscribed to them is their label from C. (If all nodes carry the same label, these are just omitted.) The box of an edge is connected to the circles of its attached nodes by lines; the attached nodes are ordered counterclockwise around the edge, starting in its north. The boxes of variables are drawn in gray. Terminal edges with two attached nodes may also be drawn as arrows from the first to the second attached node. In this case, the edge label is ascribed to the arrow.

In figures, a contextual rule $r = (L, R)$ is drawn as $L ::= R$. Small numbers above nodes indicate identities of nodes in L and R. $L ::= R_1 | R_2 \cdots$ is short for rules $L ::= R_1, L ::= R_2, \ldots$ with the same left-hand side. Subscripts "n" or "n|m$\cdots$" below the symbol $::=$ define names that are used to refer to rules in derivations, as in Figure 1 and Figure 3.

Example 1 (The Language of All Graphs). The contextual grammar in Figure 1 generates the set $\mathcal{A}$ of loop-free labeled graphs with binary edges, and Figure 2 shows a derivation with this grammar. Rules 0 and d generate $n \geqslant 0$ variables labeled with **G**; the rules n_x generate a node labeled with x, and the rules e_a insert an edge labeled with a between two nodes that are required to exist in the context.

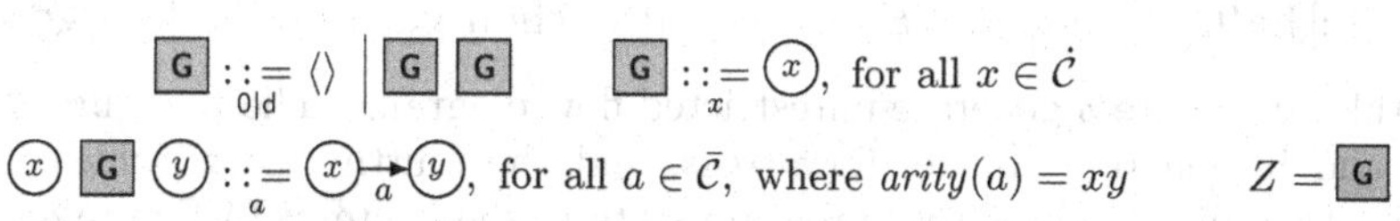

Fig. 1. A contextual grammar (generating the language of all graphs)

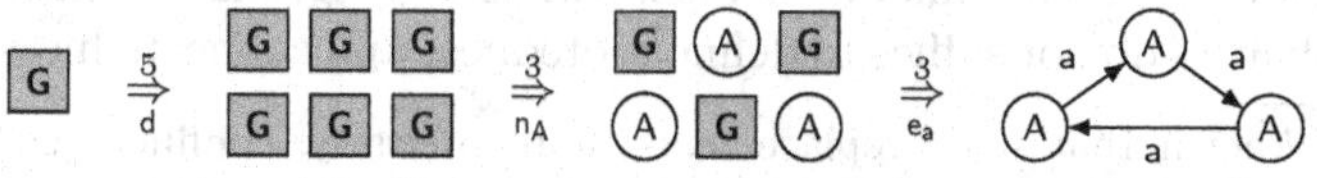

Fig. 2. A derivation with the rules in Figure 1

It is well known that the language $\mathcal{A}$ cannot be defined by hyperedge-replacement grammars [7, Chapter IV, Theorem 3.12(1)].[3] Thus, as CHR contains HR by definition, we have:

Observation 1. HR $\subsetneq$ CHR.

Flow diagrams are another example for this observation: In contrast to structured and semi-structured control flow diagrams, unrestricted control flow diagrams are not in HR, because they have unbounded tree-width [7, Chapter IV, Theorem 3.12(7)]. However, they can be generated by contextual grammars.

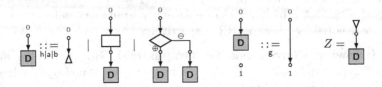

Fig. 3. Rules generating unrestricted control flow diagrams

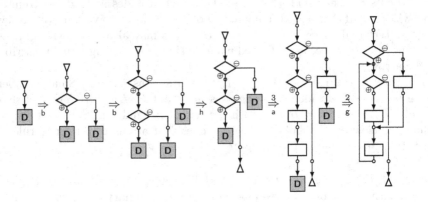

Fig. 4. A derivation of an unstructured control flow diagram

Example 2 (Control Flow Diagrams). Unrestricted control flow diagrams represent sequences of low-level instructions according to a syntax like this:

$$I ::= [\ell :] \textbf{halt} \quad | \quad [\ell :] x := E \quad | \quad [\ell_1 :] \textbf{if } E \textbf{ then goto } \ell_2 \quad | \quad [\ell_1 :] \textbf{goto } \ell_2$$

The rules in Figure 3 generate unrestricted flow diagrams. The first three rules, h, a, and b, generate control flow trees, and the fourth rule g, which is not context-free, inserts gotos to a program state in the context. In Figure 4, these rules are used to derive an "ill-structured" flow diagram.

Note that flow diagrams cannot be defined with class diagrams, because subtyping and multiplicities do not suffice to define rootedness and connectedness of graphs.

---

[3] It is well-known that node replacement (more precisely, confluent edNCE graph grammars) cannot generate $\mathcal{A}$ either [6, Thm. 4.17]. Hence, Observation 1 holds similarly for node replacement.

3 A Contextual Grammar for Statecharts

Statecharts [9] are an extension of finite automata modeling the state transitions of a system in reaction on events. The statechart in Figure 5 describes an auction. Blobs denote states, and arrows denote transitions between them. Black blobs are initial states, blobs with a black kernel are final states, and the others are inner states. Inner states may be compound, containing compartments, separated with dashed lines, which contain sub-statecharts that act independently from one another, and may themselves contain compound states, in a nested fashion. Text inside a blob names the purpose of a state or of its compartments, and labels at transitions name the event triggering them. (We consider neither more general event specifications, nor more special types of states.)

The structure of statecharts can be specified by the class diagram shown in Figure 6. The dotted circle with a gray kernel is abstract, for inner or stop states. An inner state may be composed of compartments (denoted as dashed blobs), which in turn are composed of other states (defining the sub-charts). In examples like Figure 5, a compound state is drawn as a big blob with solid lines wherein the compartments are separated by dashed lines.

The class diagram captures several structural properties of statecharts: It forbids isolated initial and final states and transitions to initial states; each compartment contains exactly one initial state, and compound states and their compartments form a tree hierarchy as the associations uniting and containing are compositions (indicated by the black diamonds at their sources).

Example 3 (A Grammar for Statecharts). The contextual rules in Figure 7 generate statecharts according to the class diagram in Figure 6. (Let us ignore the parts underlaid in gray for a moment.) The charts in these rules are drawn so that the compositions uniting and containing are just represented by drawing the blob of their target within the blob of their source. We assume (for regularity) that the topmost statechart is contained in a compartment, see the start graph Z. The rules for **S** add transitions to current states, which are either initial or inner states (drawn as gray nodes). The target of the transition is either a new final state (rule f), or a new inner state, to which further transitions may be

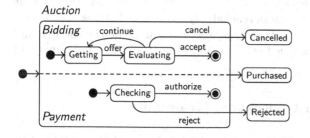

Fig. 5. A statechart modeling an auction

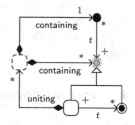

Fig. 6. A class diagram for statecharts

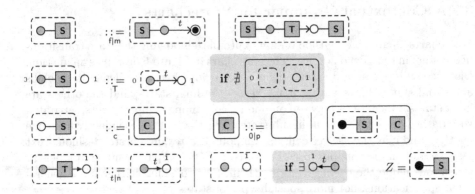

Fig. 7. Contextual rules for statecharts (with application conditions)

added (rule m), or an inner state that exists in the context, but not necessarily in the same compartment (rule T). Rule m inserts a variable named **T** that may generate a concrete transition (rule t), or leave the states separate (rule n). (This is necessary since the transitions to an inner state in a compartment need not come from a state in that compartment, like states Canceled and Rejected in Figure 5.) Finally, inner states may be filled (by rules 0 and p for the variable **C**) with compartments, each containing a statechart as in the start graph Z.

Every state in a chart should be reachable from an initial state. Reachability cannot be expressed by class diagrams alone. In order to specify this property in a meta-model, the inner state must be extended with an auxiliary attribute that determines this condition by inspecting its direct predecessors, and with a logical constraint, e.g., in OCL, which requires that the value of the attribute is true for every instance of a state.

Example 2 shows that contextual grammars can express reachability as such. In statecharts, reachability is combined with hierarchical nesting of sub-states, and cannot be specified with contextual rules. However, we may extend contextual rules with application conditions, as proposed in [11]. The parts underlaid in gray add application conditions to two rules of Figure 7. In Rule n, the condition requires that the target node of variable **T** is the target of another transition. It is easy to show that this guarantees reachability from initial states. The condition for rule T expresses yet another consistency requirement: The source and target of a transition must not lie in sister compartments of the same compound state.

4   Normal Forms of Contextual Grammars

In this section, we study the basic properties of contextual grammars. As it turns out, these properties are not fundamentally different from the properties known for the context-free case. This indicates that contextual hyperedge replacement is a modest generalization of hyperedge replacement that, to the extent one might reasonably hope for, has appropriate computational properties.

Let us first look at some normal forms of contextual grammars. We say that a restricted class C of contextual grammars is a normal form of contextual grammars (of a specified type) if, for every contextual grammar (of that type), one can effectively construct an equivalent grammar in C.

Lemma 1. *Contextual grammars in which each rule contains at most one contextual node are a normal form of contextual grammars.*

Proof. This is straightforward. Suppose we wish to implement a rule (L, R) whose left-hand side contains a variable with k attached nodes and $l \geqslant 1$ contextual nodes. We use l rules to collect the l contextual nodes one by one, finally ending up with a variable that is attached to $k + l$ nodes. The original rule is then turned into a context-free rule. □

In the context-free case, so-called epsilon rules and chain rules can easily be removed from a grammar. A similar modification is possible for contextual grammars. In this context, a rule (L, R) with $\bar{L} = \{x\}$ is an *epsilon rule* if $R = L - x$, and a chain rule if $R - y = L - x$ for a variable $y \in \bar{R}$. Note that both epsilon and chain rules are more general than in the context-free case, because L may contain contextual nodes. In particular, chain rules can make use of these contextual nodes to "move" a variable through a graph. In the case of epsilon rules, the effect of contextual nodes is that the removal of a variable is subject to the condition that certain node labels are present in the graph.

Lemma 2. *Contextual grammars with neither epsilon nor chain rules are a normal form of those contextual grammars that do not generate the empty graph.*

Proof Sketch. While the overall structure of the proof is similar to the corresponding proof for the context-free case, its details are slightly more complicated. Therefore, we give only a very rough sketch of the proof. The full proof will be given in a forthcoming extended version of this article.

The proof works as follows. First, it is shown that epsilon rules may be removed by composing non-epsilon rules with epsilon rules that remove some of the variables in the right-hand side of the original rule. Afterwards, chain rules are removed by replacing them with rules that correspond to a sequence of chain rules applied in succession, followed by the application of a non-chain rule.

The notion of composition used here has to take contextual nodes into account. Suppose we are given rules $r_1 = (L_1, R_1)$ and $r_2 = (L_2, R_2)$, such that R_1 contains a variable with the same name as the variable in L_2. We need to be able to combine both rules even if R_1 does not supply r_2 with all the necessary contextual nodes. We do this by enriching L_1 with the contextual nodes needed by r_2. However, if r_1 contains nodes (with the right labels) that are isolated in both L_1 and R_1, these are used instead rather than adding even more contextual nodes to the left-hand sides. This is a necessary precaution, because the composition of chain rules may otherwise create an infinite set of rules.

The removal of epsilon rules is non-trivial, because we have to make sure to avoid introducing deadlocks. To see this, suppose a rule r_1 creates a variable e_1 that can be removed by an epsilon rule containing a contextual node labeled a_1. Similarly, assume that r_2 creates a variable e_2 that can be removed by an epsilon rule containing a contextual node labeled a_2. Assume furthermore that r_1 generates an a_2-labeled node and r_2 generates an a_1-labeled node. Then, given a graph that contains the left-hand sides of r_1 and r_2, we can apply r_1 and r_2, followed by the epsilon rules that delete e_1 and e_2. However, if we compose r_1 with the first epsilon rule and r_2 with the second one, neither of the composed rules may be applicable, because the first contains an a_1-labeled contextual node and the second contains an a_2-labeled contextual node. Fortunately, the problem can be solved by a guess-and-verify strategy, thanks to the fact that the number of contextual nodes in the left-hand sides of rule is bounded. Roughly speaking, the guess-and-verify strategy makes sure that the required contextual nodes will be generated somewhere in the graph.

Finally, let us sketch how to remove chain rules, assuming that the grammar does not contain epsilon rules. For this, the following observation is crucial. Consider a derivation $G_0 \Rightarrow_{\mathcal{R}} G_1 \Rightarrow_{\mathcal{R}} \cdots \Rightarrow_{\mathcal{R}} G_m$ that consists of $m-1$ applications of chain rules followed by a single application of another rule. Suppose the variables replaced are $x_1, \ldots, x_m$, and let $1 \leqslant i_1 < \cdots < i_n = m$ be those indices such that $x_m = x_{i_n}$ is a direct descendant of $x_{i_{n-1}}$, which is a direct descendent of $x_{i_{n-2}}$, and so on. Then all derivation steps that replace variables in $\{x_{i_1}, \ldots, x_{i_n}\}$ can be postponed until after the other $m-n$ steps. This is because the chain rules do not create nodes that the other rules may use as contextual nodes. In other words, we can assume that $i_j = m - n + j$ for all $j \in [n]$. As a consequence, it is safe to modify Γ by adding all rules obtained by composing a sequence of chain rules with a single non-chain rule and remove all chain rules. Thanks to the observation above, the language generated stays the same.   □

Note that, unfortunately, it seems that the normal forms of the previous two lemmas cannot be achieved simultaneously.

Definition 3 (Reducedness of Contextual Grammars). In $\Gamma = \langle \mathcal{C}, \mathcal{R}, Z \rangle$, a rule $r \in \mathcal{R}$ is *useful* if there is a derivation of the form $Z \Rightarrow_{\mathcal{R}}^* G \Rightarrow_r G' \Rightarrow_{\mathcal{R}}^* H$ such that $H \in \mathcal{G}_{\mathcal{C} \setminus X}$. Γ is *reduced* if every rule in $\mathcal{R}$ is useful.

Note that, in the case of contextual grammars, usefulness of rules is not equivalent to every rule being reachable (i.e., for some G', the part of the derivation above up to G' exists) and productive (i.e., for some G, the part starting from G exists), because it is important that the pairs (G, G') are the same.

Theorem 1. *Reducedness is decidable for contextual grammars.*

Proof Sketch. Let us call a variable name ξ useful if there is a useful rule whose left-hand side variable has the name ξ. Clearly, it suffices to show that it can be decided which variable names are useful. To see this, note that we can decide reducedness by turning each derivation step into two, first a context-free step that nondeterministically "guesses" the rule to be applied and remembers the guess by relabeling the variable, and then a step using the guessed rule. Then the original rule is useful if and only if the new variable name recording the guess is useful.

Assume that the start graph is a single variable without attached nodes. Then, derivations can be represented as augmented derivation trees, where the vertices represent the rules applied. Suppose that some vertex ω represents the rule (L, R), where L contains the contextual nodes $u_1, \ldots, u_k$. Then the augmentation of ω consists in *contextual references* $(\omega_1, v_1), \ldots, (\omega_k, v_k)$, where each ω_i is another vertex of the tree, and the v_i are distinct nodes, each of which is generated by the rule at ω_i and carries the same label as u_i. The pair (ω_i, v_i) means that the contextual node u_i was matched to the node v_i generated at ω_i. Finally, in order to correspond to a valid derivation, there must be a linear order $\prec$ on the vertices of the derivation tree such that $\omega \prec \omega'$ for all children ω' of a vertex ω, and $\omega_i \prec \omega$ for each ω_i as above.[4]

Now, to keep the argument simple, assume that every rule contains at most one contextual node (see Lemma 1), and also that the label of this node differs from the labels of all nodes the variable is attached to. (The reader should easily be able to check that the proof generalizes to arbitrary contextual grammars.) The crucial observation is the following. Suppose that, for a given label $a \in \dot{\mathcal{C}}$, ω_a is the first vertex (with respect to $\prec$) that generates an a-labeled node v_a. Then, in each other vertex ω as above, if the rule contains an a-labeled contextual node u, the corresponding contextual reference (ω', v) can be replaced with (ω_a, v_a). This may affect the graph generated, but does not invalidate the derivation tree. We can do this for all vertices ω and node labels a. As a consequence, at most $|\dot{\mathcal{C}}|$ vertices of the derivation tree are targets of contextual references. Moreover, it should be obvious that, if the derivation tree is decomposed into $s(t(u))$, where the left-hand sides of the rules at the roots of t and u are the same, then $s(u)$ is a valid derivation tree, provided that no contextual references in s and u point to vertices in t. It follows that, to check whether a variable name is useful, we only have to check whether it occurs in the (finite) set of valid derivation trees such that (a) all references to nodes with the same label are equal and (b) for every decomposition of the form above, there is a contextual reference in s or u that points to a vertex in t. □

Clearly, removing all useless rules from a contextual grammar yields an equivalent reduced grammar. Thus, we can compute a reduced contextual grammar from an arbitrary one by determining the largest subset of rules such that the restriction to these rules yields a reduced contextual grammar.

Corollary 1. *Reduced contextual grammars are a normal from of contextual grammars.*

---

[4] To be precise, validity also requires that the variable replaced by the rule at ω is not attached to v_i.

By turning a grammar into a reduced one, it can furthermore be decided whether the generated language is empty (as it is empty if and only if the set of rules is empty and the start graph contains at least one variable).

Corollary 2. *For a contextual grammar Γ, it is decidable whether $\mathcal{L}(\Gamma) = \emptyset$.*

5   Limitations of Contextual Grammars

Let us now come to two results that show limitations of contextual grammars similar to the known limitations of hyperedge-replacement grammars. The first of these results is a rather straightforward consequence of Lemma 2: as in the context-free case, the languages generated by contextual grammars are in NP, and there are NP-complete ones among them.

Theorem 2. *For every contextual grammar Γ, it holds that $\mathcal{L}(\Gamma) \in$ NP. Moreover, there is a contextual grammar Γ such that $\mathcal{L}(\Gamma)$ is NP-complete.*

Proof. The second part follows from the fact that this holds even for hyperedge-replacement grammars, which are a special case of contextual grammars. For the first part, by Lemma 2, it may be assumed that Γ contains neither epsilon nor chain rules. It follows that the length of each derivation is linear in the size of the graph generated. Hence, derivations can be nondeterministically "guessed". □

It should be pointed out that the corresponding statement for hyperedge-replacement languages is actually slightly stronger than the one above, because, in this case, even the uniform membership problem is in NP (i.e., the input is (Γ, G) rather than just G). It is unclear whether a similar result can be achieved for contextual grammars, because the construction given in the proof of Lemma 2 may, in the worst case, lead to an exponential size increase of Γ.

Theorem 3. *For a graph G, let $|G|$ be either the number of nodes of G, the number of edges of G, or the sum of both. For every contextual grammar Γ, if $\mathcal{L}(\Gamma) = \{H_1, H_2, \dots\}$ with $|H_1| \leqslant |H_2| \leqslant \dots$, there is a constant k such that $|H_{i+1}| - |H_i| \leqslant k$ for all $i \in \mathbb{N}$.*

Proof Sketch. The argument is a rather standard pumping argument. Consider a contextual grammar Γ without epsilon and chain rules, such that $\mathcal{L}(\Gamma)$ is infinite. (The statement is trivial, otherwise.) Now, choose a derivation $Z = G_0 \Rightarrow G_1 \Rightarrow \dots \Rightarrow G_n$ of a graph $G_n \in \mathcal{L}(\Gamma)$, and let x_i be the variable in G_i that is replaced in $G_i \Rightarrow G_{i+1}$, for $0 \leqslant i < n$. If the derivation is sufficiently long, there are $i < j$ such that x_i and x_j have the same label and x_j is a descendant of x_i (in the usual sense). Let $i = i_1 < \dots < i_k = j$ be the indices l, $i \leqslant l \leqslant j$, such that x_l is a descendant of x_i. The steps in between those given by $i_1, \dots, i_k$ (which replace variables other than the descendants of x_i) may be necessary to create the contextual nodes that "enable" the rules applied to $x_{i_1}, \dots, x_{i_k-1}$. However, in G_j, these contextual nodes do all exist, because derivation steps do not delete nodes. This means that the sub-derivation given by the steps in which

$x_{i_1}, \ldots, x_{i_k - 1}$ are replaced can be repeated, using x_j as the starting point (and using, in each of these steps, the same contextual nodes as the original step). This pumping action can, of course, be repeated, and it increases the size of the generated graph by at most a constant each time. As there are neither epsilon nor chain rules, this constant is non-zero, which completes the proof.   □

Corollary 3. *The language of all complete graphs is not in CHR.*

6   Parsing

In [11], a parser has been briefly sketched that can be used for contextual hyperedge replacement grammars with application conditions and, therefore, for contextual grammars. The following describes the parser in more detail, including the grammar transformations that are necessary before it can be applied.

The parser adopts the idea of the *Cocke-Younger-Kasami* (CYK) parser for strings, and it requires the contextual grammar to be in *Chomsky normal form* (CNF), too. A contextual grammar is said to be in CNF if each rule is either terminal or nonterminal. The right-hand side of a terminal rule contains exactly one edge which is terminal, whereas the right-hand side of a nonterminal rule contains exactly two edges which are variables. Rules must not contain isolated nodes in their right-hand sides. In the following, we first outline that every contextual grammar Γ can be transformed into a grammar Γ' in CNF so that a parser for Γ' can be used as a parser for Γ. We then consider a contextual grammar in CNF and sketch a CYK parser for such a grammar.

If the right-hand side of a rule contains an isolated node, it is either (i) a contextual node, or (ii) a node generated by the rule, or (iii) attached to the variable of the left-hand side. In case (i), we simply remove the node from the rule. However, the parser must make sure in its second phase (see below) that the obtained rule is only applied after a node with corresponding label has been created previously. Case (ii) can be avoided if we transform the original rule set $\mathcal{R}$ to $\mathcal{R}'$ where each node generated by a rule is attached to a unary hyperedge with a new label, say $\nu \in \bar{C}$. Instead of parsing a graph G we have to parse a graph G' instead where each node is attached to such a ν-edge. Finally, case (iii) can be avoided by transforming $\mathcal{R}'$ again, obtaining $\mathcal{R}''$. The transformation process works iteratively: Assume a rule $L ::= R$ with R containing isolated nodes of kind (iii). Let $x \in \bar{L}$ with label ξ be the variable in L. This rule is replaced by a rule $L' ::= R'$ where L' and R' are obtained from L and R by removing the isolated nodes of kind (iii) and by attaching a new variable to the remaining nodes of $att(x)$, introducing a new variable name $\xi' \in X$. We now search for all rules that have ξ-variables in their right-hand sides. We copy these rules, replace all variables labeled ξ by ξ'-variables in their right-hand sides,[5] and add the obtained rules to the set of all rules. This process is repeated until no rule with isolated nodes is left. Obviously, this procedure terminates eventually.

---

[5] This procedure assumes that no rule contains more than one ξ-edge in its right-hand side. It is easily generalized to rules with multiple occurrences of ξ-edges.

We assume that the start graph is a single variable labeled ζ, for some $\zeta \in X$ with $arity(\zeta) = \varepsilon$. Thus, no ζ-edge will ever be replaced by a ζ'-edge. It is clear that $Z \Rightarrow^*_{\mathcal{R}'} G$ iff $Z \Rightarrow^*_{\mathcal{R}''} G$ for each graph $G \in \mathcal{G}_{C \setminus X}$.

Afterwards, chain rules are removed (see Lemma 2), and the obtained contextual grammar is transformed into an equivalent grammar in CNF using the same algorithm as for string grammars.[6] Based on this grammar, the parser analyzes a graph G in two phases. The first phase creates trees of rule applications bottom-up. The second phase searches for a derivation by trying to find a suitable linear order $\prec$ on the nodes of one of the derivation trees, as in the proof of Theorem 1.

In the first phase, the parser computes n sets $S_1, S_2, \ldots, S_n$ where n is the number of edges in G. Each set S_i eventually contains all graphs (called "S_i-graphs" in the following) that are isomorphic to the left-hand side of any rule, except for their contextual nodes which are left out, and that can be derived to any subgraph of G that contains exactly i edges, if any required contextual nodes are provided.

Set S_1 is built by finding each occurrence s of the right-hand side R of any terminal rule (L, R) and adding the isomorphic image s' of L to S_1, but leaving out all of its contextual nodes. Graph s' additionally points to its "child" graph s.

The remaining sets $S_i, i > 1$, are then constructed using nonterminal rules. A nonterminal rule (L, R) is reversely applied by selecting appropriate graphs s and s' in sets S_i and S_j, respectively, such that $R \cong s \cup s'$. A new graph s'' is then[7] added to the set S_k where s'' is isomorphic to L without its contextual nodes. Note that $k = i + j$ since each S_i-graph can be derived to a subgraph of G with exactly i edges. Graph s'' additionally points to its child graphs s and s'. Therefore, each instance of the start graph Z in S_n represents the root of a tree of rule applications and, therefore, a derivation candidate for G. Note that contextual nodes are not explicitly indicated in these trees because they have been removed from the S_i-graphs. Contextual nodes are rather treated as if they were generated by the rules. However, they can be easily distinguished from really generated ones by inspecting the rules used for creating the S_i-graphs.

The second parser phase tries to establish the linear order $\prec$ on the nodes of the derivation tree. The order must reflect the fact that each contextual node must have been generated earlier in the derivation. This process is similar to topological sorting, and it succeeds iff a derivation of G exists.

The run-time complexity of this parser highly depends on the grammar since the first phase computes all possible derivation trees. In bad situations, it is comparable to the exponential algorithm that simply tries all possible derivations.

---

[6] This is possible iff the $\mathcal{L}(\Gamma)$ does not contain the empty graph which is easily accomplished since chain rules have been removed.

[7] Furthermore, the parser must check whether the subgraphs of G being derivable from s and s' do not have edges in common. This is easily accomplished by associating each graph in any set S_i with the set of all edges in the derivable subgraph of G. A rule may be reversely applied to s and s' if the sets associated with s and s' are disjoint.

In "practical" cases without ambiguity (e.g., for control flow diagrams, cf. Example 2), however, the parser runs in polynomial time. Reasonably fast parsing has been demonstrated by DIAGEN [12] that uses the same kind of parser.

A simpler, more efficient way of parsing can be chosen for grammars with the following property: A contextual grammar $\Gamma = (\mathcal{C}, \mathcal{R}, Z)$ is *uniquely reductive* if its derivation relation $\Rightarrow_{\mathcal{R}}$ has an inverse relation $\Rightarrow_{\mathcal{R}^{-1}}$ (called *reduction relation*) that is is terminating and confluent. Then every graph has a reduction sequence $G \Rightarrow_{\mathcal{R}^{-1}}^{*} Y$ so that no rule of $\mathcal{R}^{-1}$ applies to Y. Confluence of reduction implies that the graph Y is unique up to isomorphism so that G is in the language of Γ if and only if Y equals Z up to isomorphism.

Let Γ be a contextual grammar with neither epsilon, nor chain rules (By Lemma 2, each contextual grammar without epsilon rules can be transformed into such a normal form). Then every right-hand side of a rule contains at least one terminal edge or one new node, and reductions $G \Rightarrow_{\mathcal{R}^{-1}}^{*} Y$ terminate, after a linear number of steps. Confluence of terminating reductions $\Rightarrow_{\mathcal{R}^{-1}}$ can be shown by checking that their *critical pairs* are strongly convergent [13]. So it can be decided whether Γ is uniquely reductive.

Since the construction of a single reduction step is polynomial for a fixed set of rules, the complexity of parsing is polynomial as well. Note, however, that parsing does not yield unique derivation structures if the reduction relation has critical pairs.

Example 4 (Parsing of Control Flow Diagrams). The grammar in Example 2 does not contain epsilon or chain rules. The right-hand sides of the rules may overlap in their interface node. Overlap in interface nodes alone does not lead to a critical pair, because the rules are still parallelly independent. The right-hand sides of the recursive rules for assignment and branching may also overlap in variables. This gives no critical pair either, because the inverse rules cannot be applied to the overlap: they violate the dangling condition. The rules are thus uniquely reductive.

7   Conclusions

In this paper we have studied fundamental properties of contextual grammars. They have useful normal forms, namely rules with at most one contextual node, grammars without epsilon and chain rules, and reduced grammars. With context-free grammars, they share certain algorithmic properties (i.e., decidability of reducedness and emptiness, as well as an NP-complete membership problem) and the linear growth of their languages. Nevertheless, contextual grammars are more powerful than context-free ones, as illustrated in Figure 8. Let NR, ASR, cCHR, and cASR denote the classes of graph languages generated by node replacement, adaptive star replacement, conditional contextual hyperedge replacement, and conditional adaptive star grammars, respectively. HR is properly included in NR [6, Section 4.3], as is NR in ASR [2, Corollary 4.9]. The proper inclusion of HR in CHR is stated in Observation 1. Corollary 3 implies that CHR neither

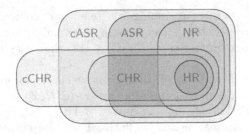

Fig. 8. Inclusion of languages studied in this paper and in [2,11]

includes NR, nor ASR, nor cCHR, because these grammars generate the language of complete graphs. We do not yet know whether ASR includes CHR; the relation of cCHR to ASR and cASR is open as well. Example 2 indicates that contextual grammars allow for a finer definition of structural properties of models than class diagrams. Application conditions do further increase this power, as discussed in Section 3.

Some work related to the concepts shown in this paper shall be mentioned here. Context-exploiting rules [4] correspond to contextual rules with a positive application condition, and are equivalent to the context-embedding rules used to define diagram languages in DIAGEN [12]. The context-sensitive hypergraph grammars discussed in [7, Chapter VIII] correspond to context-free rules with a positive application condition. We are not aware of any attempts to extend node replacement in order to define graph languages as they are discussed in this paper. The graph reduction specifications [1] mentioned in Section 6 need not use nonterminals, and their rules may delete previously generated subgraphs. They are therefore difficult to compare with contextual grammars. Example 4 shows that some contextual rules specify graph reductions, and may thus use their simple parsing algorithm. *Shape analysis* aims at specifying pointer structures in imperative programming languages (e.g., leaf-connected trees), and at verifying whether this shape is preserved by operations. Several logical formalisms have been proposed for this purpose [14]. For graph transformation rules, shape analysis has been studied for shapes defined by context-free grammars [10] and by adaptive star grammars [3]. We are currently working on shape analysis of graph transformation rules w.r.t. contextual grammars.

Future work on contextual grammars shall clarify the open questions concerning their generative power, and continue the study of contextual rules with recursive application conditions [8] that has been started in [11]. Furthermore, we aim at an improved parsing algorithm for contextual grammars that are unambiguous modulo associativity and commutativity of certain replicative rules.

Acknowledgements. We wish to thank Annegret Habel for numerous useful comments on the contents of this paper, and the reviewers for their advice to enhance the "smack of industrial relevance" of this paper.

References

1. Bakewell, A., Plump, D., Runciman, C.: Specifying pointer structures by graph reduction. Mathematical Structures in Computer Science (2011) (accepted for publication)
2. Drewes, F., Hoffmann, B., Janssens, D., Minas, M.: Adaptive star grammars and their languages. Theoretical Computer Science 411(34-36), 3090–3109 (2010)
3. Drewes, F., Hoffmann, B., Janssens, D., Minas, M., Van Eetvelde, N.: Shaped Generic Graph Transformation. In: Schürr, A., Nagl, M., Zündorf, A. (eds.) AGTIVE 2007. LNCS, vol. 5088, pp. 201–216. Springer, Heidelberg (2008)
4. Drewes, F., Hoffmann, B., Minas, M.: Context-exploiting shapes for diagram transformation. Machine Graphics and Vision 12(1), 117–132 (2003)
5. Ehrig, H., Ehrig, K., Prange, U., Taentzer, G.: Fundamentals of Algebraic Graph Transformation. EATCS Monographs on Theoretical Computer Science. Springer (2006)
6. Engelfriet, J.: Context-Free Graph Grammars. In: Handbook of Formal Languages. Beyond Words, vol. 3, ch. 3, pp. 125–213. Springer (1999)
7. Habel, A.: Hyperedge Replacement: Grammars and Languages. LNCS, vol. 643. Springer, Heidelberg (1992)
8. Habel, A., Radke, H.: Expressiveness of graph conditions with variables. In: Ermel, C., Ehrig, H., Orejas, F., Taentzer, G. (eds.) International Colloquium on Graph and Model Transformation 2010. ECEASST, vol. 30 (2010)
9. Harel, D.: On visual formalisms. Communication of the ACM 31(5), 514–530 (1988)
10. Hoffmann, B.: Shapely hierarchical graph transformation. In: Proc. of the IEEE Symposia. on Human-Centric Computing Languages and Environments, pp. 30–37. IEEE Computer Press (2001)
11. Hoffmann, B., Minas, M.: Defining models – Meta models versus graph grammars. In: Küster, J.M., Tuosto, E. (eds.) Graph Transformation and Visual Modeling Techniques 2010. ECEASST, vol. 29 (2010)
12. Minas, M.: Concepts and realization of a diagram editor generator based on hypergraph transformation. Science of Computer Programming 44(2), 157–180 (2002)
13. Plump, D.: Hypergraph Rewriting: Critical Pairs and Undecidability of Confluence. In: Sleep, M.R., Plasmeijer, M.J., van Eekelen, M.C. (eds.) Term Graph Rewriting, Theory and Practice, pp. 201–213. Wiley & Sons (1993)
14. Sagiv, M., Reps, T., Wilhelm, R.: Solving shape-analysis problems in languages with destructive updating. ACM Transactions on Programming Languages and Systems 20(1), 1–50 (1998)
15. Schürr, A., Winter, A., Zündorf, A.: The Progres Approach: Language and Environment. In: Ehrig, H., Engels, G., Kreowski, H.-J., Rosenberg, G. (eds.) Handbook of Graph Grammars and Computing by Graph Transformation. Applications, Languages, and Tools, vol. 2, ch. 13, pp. 487–550. World Scientific (1999)
16. Uesu, T.: A system of graph grammars which generates all recursively enumerable sets of labelled graphs. Tsukuba Journal of Mathematics 2, 11–26 (1978)

The Added Value of Programmed Graph Transformations – A Case Study from Software Configuration Management

Thomas Buchmann, Bernhard Westfechtel, and Sabine Winetzhammer

Lehrstuhl Angewandte Informatik 1, University of Bayreuth
D-95440 Bayreuth, Germany
firstname.lastname@uni-bayreuth.de

Abstract. Model-driven software engineering intends to increase the productivity of software engineers by replacing conventional programming with the development of executable models at a high level of abstraction. It is claimed that graph transformation rules contribute towards this goal since they provide a declarative, usually graphical specification of complex model transformations. Frequently, graph transformation rules are organized into even more complex model transformations with the help of control structures, resulting in full-fledged support for executable behavioral models.

This paper examines the added value of programmed graph transformations with the help of a case study from software configuration management. To this end, a large model is analyzed which was developed in the MOD2-SCM project over a period of several years. The model was developed in Fujaba, which provides story diagrams for programming with graph transformations. Our analysis shows that the model exhibits a strongly procedural flavor. Graph transformation rules are heavily used, but typically consist of very small patterns. Furthermore, story diagrams provide fairly low level control structures. Altogether, these findings challenge the claim that programming with graph transformations is performed at a significantly higher level of abstraction than conventional programming.

1 Introduction

Model-driven software engineering is a discipline which receives increasing attention in both research and practice. Object-oriented modeling is centered around class diagrams, which constitute the core model for the structure of a software system. From class diagrams, parts of the application code may be generated, including method bodies for elementary operations such as creation/deletion of objects and links, and modifications of attribute values. However, for user-defined operations only methods with empty bodies may be generated which have to be filled in by the programmer. Here, programmed graph transformations may provide added value for the modeler. A behavioral model for a user-defined operation may be specified by a programmed graph transformation. A model instance being composed of objects, links, and attributes is considered as an attributed graph. A graph transformation rule specifies the replacement of a subgraph in a declarative way. Since it may not be possible to model the behavior of a user-defined

A. Schürr, D. Varró, and G. Varró (Eds.): AGTIVE 2011, LNCS 7233, pp. 198–209, 2012.

operation with a single graph transformation rule, control structures are added to model composite graph transformations.

But what is the added value of programmed graph transformations? Typical arguments which have been frequently repeated in the literature in similar ways are the following ones:

1. A graph transformation rule specifies a complex transformation in a rule-based, declarative way at a much higher level of abstraction than a conventional program composing elementary graph operations.
2. Due to the graphical notation, programming with graph transformations is intuitive and results in (high-level) programs which are easy to read and understand.

This paper examines the added value of programmed graph transformations by analyzing a large model which was developed in the MOD2-SCM project (Model-Driven and Modular Software Configuration Management) [3] over a period of several years. The model was developed in Fujaba [8], which provides story diagrams for programming with graph transformations. We analyze the MOD2-SCM model both quantitatively and qualitatively to check the claims stated above.

2   MOD2-SCM

The MOD2-SCM project [3] is dedicated to the development of a model-driven product line for Software Configuration Management (SCM) systems [1]. In contrast to common SCM systems, which have their underlying models hard-wired in hand-written program code, MOD2-SCM has been designed as a modular and model-driven approach which (a) reduces the development effort by replacing coding with the creation of executable models and (b) provides a product line supporting the configuration of an SCM system from loosely coupled and reusable components.

To achieve this goal, we used Fujaba [8] to create the executable domain model of the MOD2-SCM system. The main part of the work was to (1) create a feature model [4], that captures the commonalities and variable parts within the domain software configuration management and (2) to create a highly modular domain model whose loosely coupled components can be configured to derive new products. To this end, a model library consisting of loosely coupled components that can be combined in an orthogonal way has been built around a common core.

The success of a product line heavily depends upon the fact that features that have been declared as independent from each other in the feature model are actually independent in their realizing parts of the domain model. Thus, a thorough analysis of the dependencies among the different modules is crucial [3] in order to derive valid product configurations. In order to keep track of the dependencies in large domain models, a tool based upon UML package diagrams has been developed and integrated with Fujaba to support the modeler during this tedious task [2]. In the context of the MOD2-SCM project, graph transformations were used to specify the behavior of the methods that have been declared in the domain model.

In this paper, we will discuss the added value of graph transformations especially in the development of large and highly modular software systems. The added value of Fujaba compared with other CASE tools is the ability to generate executable code out of behavioral models. Behavioral modeling in Fujaba is performed with story diagrams which are similar to interaction overview diagrams in UML2 [6]. Within story diagrams, activities and transitions are used to model control flow. Fujaba supports two kinds of activities: (1) statement activities, allowing the modeler to use source code fragments that are merged 1:1 with the generated code, and (2) story activities. A story activity contains a story pattern consisting of a graph of objects and links. A static pattern encodes a graph query, a pattern containing dynamic elements represents a graph transformation rule. Story patterns may be decorated with constraints and collaboration calls. A story activity may be marked as a "for each" activity, implying that the following activities are performed for each match of the pattern. In addition to activity nodes, story diagrams contain start, end, and decision nodes.

In the following section, we analyze the domain model of MOD2-SCM with a specific focus on story diagrams.

3   Analysis

3.1   Quantitative Analysis

Tool support was required to analyze the structure and complexity of the MOD2-SCM specification. Due to the size of the project, determining the numbers listed in Tables 1 and 2 would have been a tedious task. Therefore, we wrote a Fujaba plug-in that directly operates on Fujaba's abstract syntax graph to acquire the numbers we were interested in. We conclude from the collected numbers:

Table 1. Type and number of language elements

Model element (structural model)	Total number MOD2-SCM	Total number CodeGen2	Model element (behavioral model)	Total number MOD2-SCM	Total number CodeGen2
Packages	68	18	Story diagrams	540	339
Classes	175	162	Story patterns	988	850
Abstract Classes	18	28	Objects	1688	1997
Interfaces	32	10	Negative objects	42	22
Attributes	177	181	Multi-objects	25	9
Methods	650	443	Links	725	1121
Generalizations	220	247	Paths	13	7
Associations	148	166	Statement activities	264	64
			Collaboration calls	1183	711
			For each activities	27	88

1. The number of story patterns per story diagram is rather low (an average of 1.83 story patterns were used per story diagram). This indicates a procedural style of the model and methods of lower complexity.

Table 2. Significant ratios of language elements

Metric	MOD2-SCM	CodeGen2	Metric	MOD2-SCM	CodeGen2
Classes/package	2.57	9	Negative objects/pattern	0.04	0.03
Attributes/class	1.01	1.12	Multi-objects/pattern	0.03	0.01
Methods/class	3.71	2.73	Paths/pattern	0.01	0.01
Patterns/story diagram	1.83	2.51	Statement activities/story pattern	0.27	0.08
Objects/pattern	1.71	2.35	Collaboration calls/story pattern	1.20	0.84
Links/pattern	0.73	1.32			

2. Within a story pattern, only a few objects and links are used (1.71 objects and 0.73 links, respectively).
3. The number of collaboration calls is rather high (an average number of 1.2 collaboration calls per pattern was determined).
4. Given the fact that we tried to use story patterns as much as possible, the fraction of statement activities is still rather high (0.27 statement activities per story activity). Furthermore, the seamless integration of hand-written Java code and story patterns provides lots of advantages, but it also implies one big disadvantage: The model is no longer platform-independent. For example, changing the code generation templates to generate C# code is no longer possible without manually changing each statement activity in the domain model as well.
5. Complex elements within story patterns (negative objects, multi-objects or paths) were used only very rarely within the domain model.

Generally, the specification of the MOD2-SCM system is highly procedural and essentially lies at the same level of abstraction as a conventional program. The average complexity of the implemented methods is rather low. Furthermore, the graph transformation rules are mostly limited to finding simple patterns in the object graph and to inserting a single new object at the appropriate position and/or calling another method.

3.2 Qualitative Analysis

In addition to interpreting the numbers shown in the previous section, we took a closer look at the story diagrams of the domain model to examine the expressive power of the modeling language and the readability of story diagrams. With respect to story patterns, we observed:

6. Story patterns are easy to read due to the graphical notation. Furthermore, story patterns potentially have a high expressive power, but this power is only rarely exploited since the highly modular architecture results in a fine-grained decomposition of the domain model.

We drew the following conclusions concerning control flow:

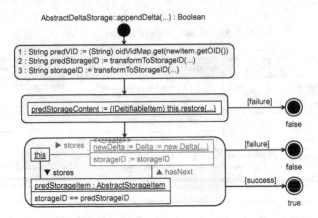

Fig. 1. Story diagram of average complexity

7. It is evident that Fujaba does not provide any "higher"-level control structures. In fact, in terms of control structures, the level of abstraction provided by Fujaba lies even below conventional programming languages, as story diagrams are very similar to flow charts.

8. In many cases, Java statement activities are mixed up with story patterns, e.g., for exception handling. Fujaba does not catch external exceptions raised in the execution of story patterns, e.g., by external Java methods executed via collaboration calls. Such low-level details need to be handled in Java.

9. Furthermore, Fujaba itself does not provide any mechanisms to iterate over standard collections. Collections are implicitly used within for-each activities, but no explicit support is provided for the user.

10. The graphical programming style may result in loss of the overall picture if the diagrams are too large and complex. Readability suffers especially (but not only) when the story diagram does not fit onto a single screen.

4   Examples

In this section, we give some examples that reinforce the statements of the previous section. Figure 1 represents a method implementation which we consider to be of *average complexity* (according to the collected metrics data). It is a typical example for the observations (1) – (5) made in Section 3.1 as well as observation (6). The method is used to append a forward delta in order to store a new state of a versioned object. The first story pattern consists only of collaboration calls that retrieve different required parameters. The second pattern retrieves the content of the predecessor version (which must have been stored as a delta). This content is used in the third pattern (the only graph transformation rule) to compute the difference and to store it at the appropriate position in the object graph.

All story patterns occurring in Figure 1 are quite simple. Figure 2 shows a *(moderately) complex story pattern* which stores a backward delta. The enclosing story diagram is called when a version stored as baseline is to be replaced with a successor

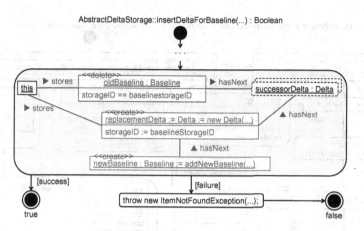

Fig. 2. A (moderately) complex story pattern

version. To this end, the old baseline is replaced with a new baseline and a backward delta. In fact this is one of the most complex patterns throughout the whole MOD2-SCM model, which does not include any patterns with more than 6 objects. This example illustrates a strength of story patterns: The effect of a graph transformation is documented in an easily comprehensible way. However, most patterns are much simpler, implying that the equivalent Java code would as well be easy to grasp (observation (6)).

Figure 3 shows the story diagram with the *highest number of story patterns* throughout the MOD2-SCM project. The story diagram is used to update the local version information within the user's workspace after changes have been committed to the server. This method is very complex if we take the average number of patterns per story diagram into account. However, its individual steps are not complex at all. Listing 1 shows the purely hand-written implementation in Java. Essentially, the method consists of a single loop iterating over a list of object identifiers. Two of the story patterns contain plain Java code since Fujaba does not supply high level constructs for iterating over standard collections (observation (9)). The remaining story patterns are also very simple (observation (5)). Thus, it is not a big challenge to code this story diagram in Java.

The story diagrams presented so far do not contain statement activities, except for a small activity in Figure 2. We used statement activities only when they were impossible or awkward to avoid. Observation (4) showed that our attempt to eliminate story patterns was only successful to a limited extent. The next two examples demonstrate the reasons for that.

The first example (Figure 4) shows a story diagram which is used to configure the MOD2-SCM repository server at runtime according to the features selected by the user. This method is *inherently procedural* and consists of a large number of conditional statements for handling the different cases. The modeler decided to code the method body as a single statement activity. Splitting this activity up into many decision nodes and activity nodes containing small code fragments would have resulted in a huge and unreadable diagram (observation (7)).

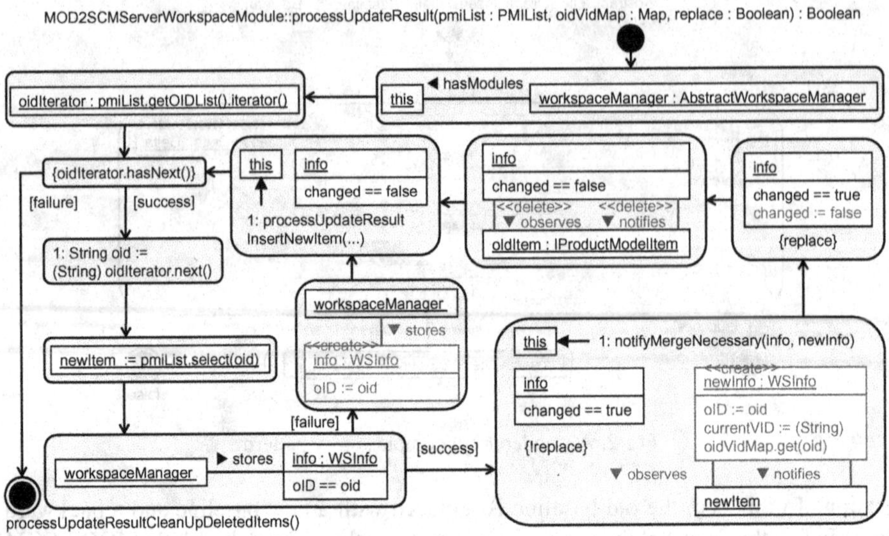

Fig. 3. Story diagram that contains a high number of patterns

Listing 1. Manual implementation of the method shown in Figure 3

```
1  public boolean processUpdateResult(PMIList pmiList, Map<String, String>
       oidVidMap, boolean replace) {
2    Iterator oidIterator = pmiList.getOIDList().iterator();
3    while (oidIterator.hasNext()) {
4      String oid = (String) oidIterator.next();
5      IProductModelItem newItem = pmiList.select(oid);
6      WSInfo info = getWorkspaceManager().getWSInfos().get(oid);
7      if (info != null) {
8        IProductModelItem oldItem = info.getItem();
9        if (!replace && info.isChanged()) {
10         WSInfo newInfo = new WSInfo(oid, (String) oidVidMap.get(oid));
11         newInfo.setItem(newItem);
12         newInfo.setObservable(newItem);
13         notifyMergeNecessary(info, newInfo);
14       }
15       if (replace && info.isChanged())
16         info.setChanged(false);
17       if (!info.isChanged() && oldItem != null) {
18         info.setItem(null);
19         info.setObservable(null);
20       }
21     } else {
22       info = new WSInfo(oid);
23       getWorkspaceManager().addWSInfo(info);
24     }
25     if (!info.isChanged())
26       processUpdateResultInsertNewItem(oldItem, newItem, info, oidVidMap);
27   }
28   processUpdateResultCleanUpDeletedItems();
29 }
```

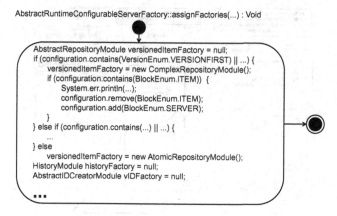

AbstractRuntimeConfigurableServerFactory::assignFactories(...) : Void

```
AbstractRepositoryModule versionedItemFactory = null;
if (configuration.contains(VersionEnum.VERSIONFIRST) || ...) {
    versionedItemFactory = new ComplexRepositoryModule();
    if (configuration.contains(BlockEnum.ITEM)) {
        System.err.println(...);
        configuration.remove(BlockEnum.ITEM);
        configuration.add(BlockEnum.SERVER);
    }
} else if (configuration.contains(...) || ...) {
    ...
} else
    versionedItemFactory = new AtomicRepositoryModule();
HistoryModule historyFactory = null;
AbstractIDCreatorModule vIDFactory = null;
```

Fig. 4. Story diagram that represents a highly procedural example

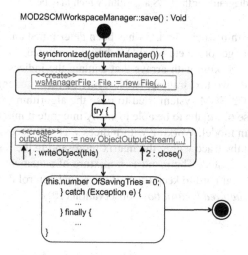

MOD2SCMWorkspaceManager::save() : Void

synchronized(getItemManager()) {

<<create>>
wsManagerFile : File := new File(...)

try {

<<create>>
outputStream := new ObjectOutputStream(...)

↑1 : writeObject(this)   ↑2 : close()

this.number OfSavingTries = 0;
        } catch (Exception e) {
            ...
        } finally {
            ...
}

Fig. 5. Story diagram that illustrates the JSP syndrome

Figure 5 is a good example how hand-written code fragments are placed around story patterns (observation (8)). The story diagram is used to save the state of the workspace manager. To ensure that the execution is synchronized, the body is embedded into a synchronization statement. Furthermore, if the write operation fails, the save method is re-executed (until the maximum number of tries is exceeded). Again, this method implementation is highly procedural and story driven modeling does not seem to be the most appropriate formalism for this task. The story diagram is written in a *JSP-like style*, including statement activities containing fragments of Java text which do not even correspond to complete syntactical units.

Figure 6 shows a story diagram that is used to calculate differences on text files based upon the well-known Longest Common Subsequence (LCS) algorithm. Figure 6 depicts

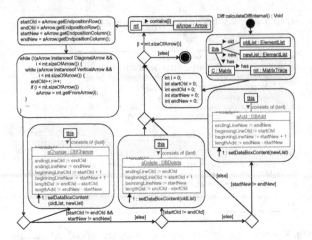

Fig. 6. Story diagram for the LCS algorithm which may be coded easily in Java

the part of the algorithm where the LCS has been determined and the difference (represented by add, change, or delete blocks) is being computed. The original algorithm performs best when working with arrays that contain the indices of the longest common subsequences in two text fragments. Nevertheless, the student who implemented this part of the MOD2-SCM system tried to raise the algorithm to the object-oriented level and to make use of Fujaba to be able to easily integrate it into the already existing MOD2-SCM domain model. Working with indices was still necessary, though objects for the matrices and the trace within the matrix representing the longest common subsequence have been created. The result is a mixture of statement activities and story patterns, which makes it hard to keep track of the actual control flow within the shown story diagram (*unstructured control flow*, observation (10)).

5   Discussion

5.1   Results from the MOD2-SCM Project

In this paper, we examined the added value of programmed graph transformations with the help of a case study from software configuration management. Based on this case study, may we convince hard-core Java programmers to program with graph transformations instead? The examination of the story diagrams developed in the MOD2-SCM project suggests the answer "no". Altogether, story diagrams are written in a strongly procedural style at a level of abstraction which hardly goes beyond Java and is even partially located below Java or other current programming languages.

The authors of the MOD2-SCM model made extensive use of story patterns. However, our quantitative analysis showed that story patterns are typically composed of a very small number of objects and links. Furthermore, advanced features such as negative objects/links, multi-objects, and paths are only very rarely used. Altogether, the potential of story patterns - the declarative specification of complex graph transformations - is only exploited to a severely limited extent.

As far as story patterns are concerned, the graphical notation is intuitive and enhances readability, in particular in the case of more complex graph transformations. With respect to control flow, however, the graphical notation may have a negative impact on readability. Essentially, story diagrams are conventional flow charts, which are well known for the "goto considered harmful" syndrome. Control structures known from structured programming are missing. In this respect, story diagrams fall behind conventional programming languages such as Java.

5.2   Generalization of Results

Let us summarize the most important general observations derived from the case study:

1. The behavioral model is highly procedural.
2. The expressive power of graph transformation rules (story patterns) is hardly exploited.

It might be argued that these findings are specific to the case study since providing a product line requires the fine-grained decomposition of the overall domain model into a set of rather small reusable components. However, this style of development is not only applied to product lines, but it should anyhow be applied in any large development project. To check this assumption, we ran our metrics tool on Fujaba's CodeGen2 model (the bootstrapped Fujaba code generator). The results were very similar to the data collected from the MOD2-SCM project (see again Tables 1 and 2).

It could also be argued that the authors of the story diagrams lacked the expertise to fully exploit the features of the modeling language. Although some minor parts of the MOD2-SCM project were developed by students who did not have much experience in programming with graph transformations, the biggest part of the system was implemented by experienced Fujaba modelers. Furthermore, the analysis of CodeGen2, which was developed by the authors of Fujaba themselves, yielded similar results.

Finally, it might be argued that the procedural style of the Fujaba models is due to the modeling language. However, this argument does not explain the fact that advanced features of story patterns such as paths, constraints, negative objects and links, and multi-objects were only rarely used. Nevertheless, we decided to examine a large specification written in another language to check this argument. The specification was developed in a Ph.D. thesis in the ECARES project, which was concerned with reverse engineering and reengineering of telecommunication systems [5]. The specification was written in PROGRES [7], a language for programming with graph transformations providing many advanced features (multiple inheritance, derived attributes and relations, object-orientation, genericity, graph transformation rules with similar features as in Fujaba, high-level control structures, backtracking, and atomic transactions).

The data displayed in Tables 3 and 4 were collected from the complete specification, as developed in the Ph.D. thesis by Marburger. By and large, the results are consistent with the metrics data collected from the Fujaba models:

1. The ECARES specification has a strongly procedural flavor. This is indicated by the ratio of the number of graph tests and graph transformation rules related to the number of programmed graph queries and transactions: There are twice as many programmed methods as elementary graph tests and transformation rules.

Table 3. Type and number of language elements in ECARES

Model element	Total number	Model element	Total number
Packages	21	Optional nodes	9
Node classes or types	190	Set nodes	40
Generalizations	193	Edges	374
Intrinsic attributes	87	Negative nodes and edges	13
Derived attributes	9	Positive and negative paths	92
Meta attributes	8	Transactions (update methods)	299
Edge types	21	Queries	8
Textual path declarations	32	Assignments	659
Graphical path declarations	40	Calls	684
Graph tests	55	Sequences	335
Graph transformation rules	92	Conditional statements	270
Mandatory nodes	559	Loops	83

Table 4. Significant ratios of language elements in ECARES

Metric	Value	Metric	Value
(Classes + types)/package	9.05	(Negative nodes + edges)/graphical definitions	0.07
Attributes/(classes + types)	0.55	Set nodes/graphical definitions	0.21
(Graph tests + graph transformation rules + queries + transactions)/(classes + types)	2.39	(Positive + negative paths)/graphical definitions	0.49
(Graph tests + graph transformation rules)/(queries + transactions)	0.48	(Assignments + calls)/(queries + transactions)	4.37
Nodes/graphical definitions	2.99	Control structures/(queries + transactions)	2.24
Edges/graphical definitions	2.00		

2. Graphical definitions (graph tests, graph transformation rules, and graphical path declarations) are rather small. On average, a graphical definition contains about 3 nodes and 2 edges. These numbers are a bit larger than in MOD2-SCM and Code-Gen2. However, it has to be taken into account that relationships are represented in ECARES always as nodes and adjacent edges. Thus, a graph transformation rule for inserting a relationship requires at least 3 nodes and 2 edges. In the publications on ECARES, considerably more complex rules were selected for presentation, but these rules are not representative.

3. The data differ with respect to the utilization of advanced features in graphical definitions. In particular, paths are used in about 50% of all graphical definitions. Eliminating paths would result in larger graphical definitions. Thus, altogether the graphical definitions are slightly more complex than in the studied Fujaba models.

6   Conclusion

We investigated the added value of programmed graph transformations with the help of a large case study from software configuration management. Our analysis showed that the model developed in the MOD2-SCM project exhibits a strongly procedural flavor. Furthermore, graph transformation rules are heavily used, but consist typically of very small and simple patterns. Finally, we have reinforced our findings with metrics data collected from other projects utilizing programmed graph transformations.

Examining a few large models does not suffice to evaluate the added value of programmed graph transformations. However, our analysis indicates that the level of abstraction is not raised as significantly as expected in comparison to conventional programming. In the models we studied, the modeling problem seems to demand for a procedural solution. Furthermore, modularization of a large model may result in a fine-grained decomposition such that each method only has to deal with small patterns and has to provide a small piece of functionality. Further case studies are required to check whether these effects also apply to other applications.

References

1. Buchmann, T., Dotor, A.: Towards a model-driven product line for SCM systems. In: McGregor, J.D., Muthig, D. (eds.) Proc. of the 13th Int. Software Product Line Conference, vol. 2, pp. 174–181. SEI (2009)
2. Buchmann, T., Dotor, A., Klinke, M.: Supporting modeling in the large in Fujaba. In: van Gorp, P. (ed.) Proc. of the 7th International Fujaba Days, pp. 59–63 (2009)
3. Dotor, A.: Entwurf und Modellierung einer Produktlinie von Software-Konfigurations-Management-Systemen. Ph.D. thesis, University of Bayreuth (2011)
4. Kang, K.C., Cohen, S.G., Hess, J.A., Novak, W.E., Peterson, A.S.: Feature-oriented domain analysis (FODA) feasibility study. Tech. rep., Carnegie-Mellon University Software Engineering Institute (1990)
5. Marburger, A.: Reverse Engineering of Complex Legacy Telecommunication Systems. Berichte aus der Informatik. Shaker-Verlag (2005)
6. OMG: OMG Unified Modeling Language (OMG UML), Superstructure V2.2 (version 2.2) (February 2009)
7. Schürr, A., Winter, A., Zündorf, A.: The PROGRES Approach: Language and Environment. In: Handbook of Graph Grammars and Computing by Graph Transformation, vol. 2: Applications, Languages, and Tools, pp. 487–550. World Scientific (1999)
8. Zündorf, A.: Rigorous object oriented software development. Tech. rep., University of Paderborn, Germany (2001)

A Case Study Based Comparison
of ATL and SDM

Sven Patzina and Lars Patzina

Center for Advanced Security Research Darmstadt (CASED), Germany
{sven.patzina,lars.patzina}@cased.de

Abstract. In model driven engineering (MDE) model-to-model transformations play an important role. Nowadays, many model transformation languages for different purposes and with different formal foundations
have emerged. In this paper, we present a case study that compares the Atlas Transformation Language (ATL) with Story Driven Modeling (SDM)
by focusing on a complex transformation in the security domain. Additionally, we highlight the differences and shortcomings revealed by this
case study and propose concepts that are missing in both languages.

Keywords: Atlas Transformation Language, Story Driven Modeling,
Live Sequence Charts, Monitor Petri nets, transformation.

1   Introduction

Model-driven engineering (MDE) demands model-to-model transformations between models on different abstraction levels. Based on this idea, a model-based
development process for security monitors [15] is developed that allows for the
abstract specification and automated generation of correct security monitors in
software (C, Java) and hardware (VHDL, Verilog). Specifications are modeled as
use and misuse cases with extended Live Sequence Charts (LSCs). Due to the expressiveness of LSCs, the process foresees a more explicit intermediate language
– the Monitor Petri nets (MPNs), a Petri net dialect with special start and end
places and deterministic execution semantics. This more explicit representation
with a less complex syntax is easier to process than the LSCs itself.

In this context, a rule-based model-to-model transformation language is intended for the complex step from LSCs to MPNs, because a rule-based approach
seems to be less error-prone compared to a manual implementation of the pattern matching process for each transformation in a general-purpose programming language. Nowadays, various transformation languages have emerged with
a different purpose, feature set, and formal foundation. On one side, there are
languages that are based on graph grammar theory such as SDM [6] and on the
other side, languages such as ATL [9], whose semantics has been formalized by
using e.g., abstract state machines [5] and rewriting logics [18]. In contrast to
existing comparisons that use classical examples [4] or more complex examples
by focusing on special properties such as inheritance [21], this case study differs
in the application domain.

A. Schürr, D. Varró, and G. Varró (Eds.): AGTIVE 2011, LNCS 7233, pp. 210–221, 2012.
© Springer-Verlag Berlin Heidelberg 2012

In this paper, we show the differences, advantages and disadvantages of the Atlas Transformation Language (ATL) in version 3.1 and Story Driven Modeling (SDM) on an example of a real-world transformation. Our main contributions are the application of two transformation languages to a complex, real-world example in the security domain and specific proposals for extending the concepts of the transformation languages.

In the following, Sec. 2 introduces the transformation scenario. Then Sec. 3 presents the requirements and analyses appropriate languages for our purpose – ATL and SDM. In Sec. 4, selected rules of the transformation in ATL and SDM that show concepts and differences are compared, and the evaluation is described. The result of the comparison and suggestions for additional features for the two transformation languages are shown in Sec. 5. Section 6 concludes the paper.

2   Running Example

In this section, a Car-to-Infrastucture scenario, where a `car` communicates with a `tollbridge`, will be presented and used in the remainder of this paper. The example is based on metamodels that are reduced versions of those used in the case study, depicted as Ecore/MOF diagrams in Figure 1a) and b). Figure 1c) shows a communication protocol (use case) in concrete LSC syntax and Figure 1d) the corresponding MPN representation.

LSCs are an extension of Message Sequence Charts that in addition offer a distinction between *hot* and *cold* elements. Thereby, hot elements are mandatory and have to occur and cold elements are optional. Additionally, LSCs can have two forms, a *universal LSC* with a *prechart* (precondition) before the *mainchart* or an *existential LSC* without a precondition.

The LSC in Figure 1c) shows an existential LSC that models the exchange of asynchronous messages between LSC objects in a mainchart. When the car approaches the tollbridge, it sends a `connect()` message to the tollbridge that is represented as hot message. The tollbridge acknowledges this message with an `ack()`. After receiving this message, the car has to send some information `data_a()`. Then the car is allowed to send additional information `data_b()`, modeled as cold message. The communication has to be terminated by the car by sending a `disconnect()` message.

The metamodel of the *LSC* diagrams is depicted in Fig. 1a), where *LSC*s and the derived *ExistentialLSC*s are contained in the root class *LSCDocument*. Furthermore, *ExistentialLSC*s contain *LSCObject*s and a *Mainchart*. A *Message* starts and ends on a *Location* that is contained in the related *LSCObject* as an ordered set.

Similar to *LSC*s, *MPN*s have an *MPNDocument* as root class, depicted in Figure 1b). The MPN, derived from the use case LSC, is represented in Figure 1d). This MPN is a more operational description of the behavior expressed by the LSC. *MPN*s are composed of different kinds of places (depicted as circles), transitions (realized as black bars), and arcs that connect places and transitions.

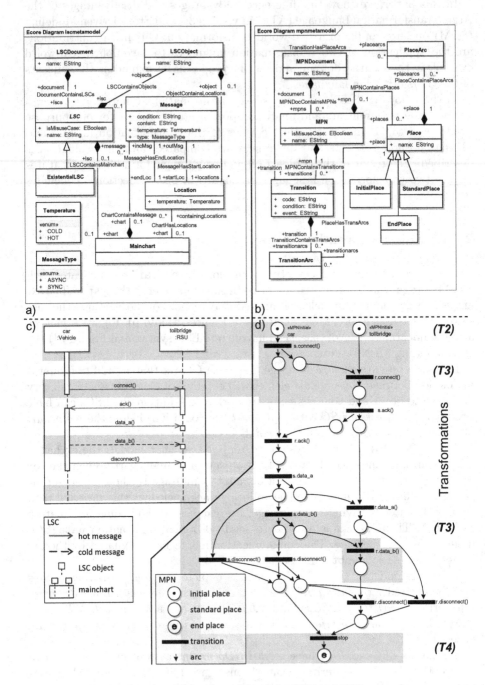

Fig. 1. Metamodels and models of the LSC-to-MPN example

The transformation from LSCs to MPNs basically consists of four transformation steps, highlighted in grey in the concrete example. *(T1)* It starts with the creation of an MPN for each LSC. *(T2)* Each *LSC object* is represented as an *initial place* annotated with the name of the object. *(T3)* A *hot asynchronous message* such as the connect() message is split in one *transition* for the sending and one for the receiving *event* and three *standard places*. These are connected with *arcs* where one place is on the sender side, one is on the receiver side, and one secures the order of sending and receiving of the message. This pattern is also used for *cold messages* with additional bypass-transitions that realize the optional nature of cold messages. *(T4)* The pattern is finalized by an *end place* where all possible paths through the MPN are synchronized by transitions with the *event* "stop".

3 Related Work

In the last years, publications about comparisons between different transformation languages evolved. On the one side, there is the Transformation Tool Contest[1] event series, where solutions for special transformation problems are submitted and compared. As [2] stated, this contest can be used as source for insights of the strong and weak points of transformation tools, but has no clear focus on achieving really comparable results. On the other side, publications cope with small classical examples such as UML2RDBMS [4], concentrate only on graph-based transformation languages [17] or focus on a small subset of properties such as inheritance [21]. The transformation from extended LSCs to the corresponding MPN representation is in contrast to the afore mentioned comparisons based on a complex, real-world example located in the security domain.

Based on the transformation steps (*T1* to *T4*) that are derived from the example in Section 2 the following requirements can be formulated:

R1) 1-to-1 (Model). For each LSC diagram, a single MPN should be generated. Therefore, for each source model one target model is created. *(T1)*

R2) m-to-n (Element). One or more elements in the source model have to be mapped to one or more elements in the target model. *(T3)*

R3) Traceability Links. For processing optional elements, traceability links are needed to be able to add bypass-transitions in the target model. *(T3)*

R4) Attributes. The language must be able to handle attributes of model elements. It has to check and generate attributes in the target model. *(T2)*

R5) In-place. For optimizations of the target model, some kind of in-place transformation on the target model is required. *(T4)*

R6) Deletion. For post-processing, it is necessary to delete elements from the target model to remove redundant places and unnecessary transitions. *(T4)*

R7) Recursive Rules. For the synchronization at the end of an LSC, with an unknown number of places and LSC objects during specification, recursive operations on model elements are needed. *(T4)*

---

[1] http://planet-research20.org/ttc2011/

Table 1. Requirements for the LSC-to-MPN transformation

Req.	ATL [9]	ETL [11]	QVTo [14]	PROGRES [16]	SDM [6]	TGG [10]	VIATRA2 [20]
R1)	✓	✓	✓	✓[1]	✓[1]	✓	✓
R2)	✓	−[2]	✓	✓	✓	✓	✓
R3)	✓	o[3]	✓	o[3]	o[3]	✓	o[3]
R4)	✓	✓	✓	✓	✓	−	✓
R5)	o[4]	✓	o[4]	✓	✓	−	✓
R6)	o[5]	✓[6]	✓	✓	✓	−	✓
R7)	o[7]	✓	✓	✓[8]	✓[8]	−	✓
R8)	✓	✓	✓	✓	✓	o[9]	✓
R9)	✓	✓	✓	−	✓	✓	✓

✓fulfilled; o partly fulfilled; - not fulfilled; [1]in-place; [2]only 1-to-n; [3]manual; [4]refining mode; [5]new transformation; [6]with EOL; [7]as helper; [8]control flow and path expressions; [9]bidirectional

R8) Unidirectional. Some elements of the LSC have no bijective mapping. A fixed loop of n iterations is, e.g., unwound to n representations of its content. *(T1)*

R9) Tool Support. For the realization of the development process, a reliable implementation of the transformation language is needed.

Based on these requirements, Table 1 compares state-of-the-art rule-based model transformation languages. While PROGRES, SDM, TGG, and VIATRA2 are based on graph transformation (GT) principles, ATL, ETL, and Operational QVT (QVTo) are not formalized. First approaches for ATL are made using abstract state machines [5] and rewriting logics [18]. The introduced GT languages, excluding TGGs, are hybrid and support the modeling of control flow. In contrast to all other approaches, TGGs are fully declarative and bidirectional transformations have to be specified as mappings of source and target elements simultaneously. This hampers the specification of the rules, because no bijective mapping as stated in R8 exists, and R7 is not supported. So TGG does not fit to the requirements.

There are two groups, on one hand ATL, ETL, and QVTo and on the other hand PROGRES, SDM, and VIATRA2. Because many comparisons between languages within one of the groups exist, e.g., [8,17], one language from each group is chosen.

ATL, the commonly used model transformation language in the Eclipse community and SDM that is used in the meta-CASE tool MOFLON [1] will be compared because of their differences. In SDM the rules, embedded in activities of an activity diagram, are scheduled by an explicitly modeled control flow. Contrary to SDM, the rules of an ATL 3.1 transformation are conditioned by OCL expressions and functionality can be delegated to helper functions. Using these concepts the execution sequence is derived from implicit relations between the rules.

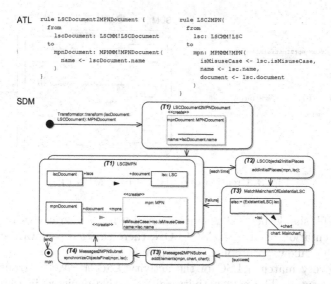

Fig. 2. Initial rules of the transformation (T1)

4   Comparison of the Transformations

In this section, rules of the LSC-to-MPN transformation are presented that show
commonalities and differences of ATL and SDM. Thereby, the requirements derived above that are not satisfied by one or both languages (R3, R5, R6 and R7)
are examined and additional missing features are suggested. After that, the implementation of the transformations is evaluated.

4.1   Transformations

The SDM part of Figure 2 maps the basic steps of the transformation, derived in
Section 2 to the activities of the SDM: *(T1) LSCDocument* to *MPNDocument*
and *LSC* to *MPN*, *(T2) LSCObjects* to *InitialPlaces*, *(T3) Messages* to an *MPN*
subnet, and *(T4)* the synchronization to *EndPlaces* in *MPN*.

(T1) The first transformation rules in Figure 2 translate the *LSCDocument*
with its *LSC*s to an *MPNDocument* with corresponding *MPN*s. The ATL rule,
LSCDocument2MPNDocument, corresponds to the first activity in the SDM. In
the SDM rule, *lscDocument* is already bound as a parameter and an *MPNDocument* is created in the first activity. Contrary to the SDM, the ATL rule needs
no already matched (bound) objects for the navigation. So an *LSCDocument* is
matched in the *from* part (left side) and an *MPNDocument* is created in the *to*
part (right side) of the rule. The second rule *LSC2MPN* is very similar in both
languages. In the SDM, the activity around the pattern is a *foreach*-activity that
uses the bound *lscDocument* to find all *LSC*s and adds for each *lsc* an MPN
with the same attributes to the *mpnDocument*.

These first transformation rules already reveal the main difference of the two
languages. While SDM explicitly relies on a control flow between the declarative

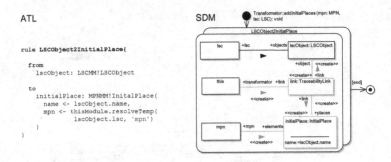

Fig. 3. Creation of initial places of MPN (T2)

patterns, ATL should be used as long as possible in a declarative way [9]. Beside the implementation of SDM in MOFLON, there exists a backward compatible extension that allows for implicit rule scheduling [13].

(T2) For every matched LSC in the activity *LSC2MPN*, the transformation steps are executed. The next activity calls the rule depicted in Figure 3 that translates all *LSCObjects* to *InitialPlaces* of the MPN. The ATL rule matches every *LSCObject* specified in the *from* part of the rule, which corresponds to the unbound object *lscObject* in the SDM. The *to* part of the ATL rule corresponds to the pattern at the bottom of the SDM activity.

Here, another difference between ATL and SDM emerges. In ATL each application of a rule automatically produces traceability links between the matched source and the created target elements. Such a mechanism does not exist in standard SDM. Therefore, an additional metamodel containing the transformation rules as operations and constructs for the management of traceability links (*TraceabilityLink*) has to be explicitly modeled as shown in [7]. Traceability links in ATL have to be used for adding a reference to the already created MPN. This is done by the predefined helper `resolveTemp` and name matching. As parameters the elements of the source model and the name of the element in the target model, defined in the rule that has matched the source elements, are passed.

(T3) When defining more complex rules such as the transformation of asynchronous messages, two different approaches have to be used. In SDM, enabled by the explicit control flow modeling, the transformation can be defined in one rule, whereas, in ATL three rules have to be specified.

For the SDM in Figure 4, the manually added traceability links are used to identify all places in the MPN that have not been synchronized via a hot message in the LSC. In the activity *CreatePlacesForAsyncMessage* three *StandardPlaces* for the LSC message are created. The following *foreach*-activity *CreatePlaces-ForStartLocation* generates a transition with corresponding arcs for every place that has a link to the source of the message (*LSCObject*). This includes the bypass-transitions for cold messages. The *statement*-activity *TraceabilityLink-ForStartLocation* calls an SDM method that manages the traceability links. The second part of the SDM, which is collapsed, performs the similar transformation for the target of the message.

SDM

Transformator::addAsyncMessage (mpn: MPN, startObject: LSCObject, endObject: LSCObject, message: Message): void

CreatePlacesForAsyncMessage

<<create>>
sPA: StandardPlace   +places   +mpn   mpn
                      <<create>> +mpn

                                  +mpn
<<create>> +places   <<create>> <<create>> +places
sPS: StandardPlace   <<create>> sPB: StandardPlace

TraceabilityLinksForStartLocation
addLocationPlace(message.getStartLocation(), link, place, sPA);

[each time]

ConnectPlacesForStartLocation

this   +transformator   +link   link: TraceabilityLink   +link   +object   startObject   sPS

+link   +link   +place

+places   <<create>>   <<create>>   <<create>>
place: Place   +place   +placearcs   pAA: PlaceArc   tAS: TransitionArc   +transitionarcs   sPA

          <<create>>   +placearcs   <<create>>   +transitionarcs   +place
                                                                                        <<create>>
                        +transition   <<create>>
          <<create>>   +transition   +transition   +transitionarcs   <<create>>
mpn   +mpn   +transitions   tA: Transition   +transition   +transitionarcs   tAA: TransitionArc
          <<create>>   event:="s."+message.getContent()   <<create>>

[end]

ConnectLocationsForEndLocations

[end]   [each time]

TraceabilityLinksForEndLocation
addLocationPlace(message.getEndLocation(), link, place, sPB);

ATL

```
rule AsyncMPNPattern{
  from
    lscAM : LSCMM!Message(lscAM.isMessageAsync)
  to
    sPA : MPNMM!StandardPlace(
      mpn <- thisModule.resolveTemp(lscAM.endLocation.object.lsc, 'mpn') ),
    sPS : MPNMM!StandardPlace( placeArcs <- tAS,
      mpn <- thisModule.resolveTemp(lscAM.endLocation.object.lsc, 'mpn') ),
    sPB : MPNMM!StandardPlace( placeArcs <- tAB,
      mpn <- thisModule.resolveTemp(lscAM.endLocation.object.lsc, 'mpn') ),
    tA : MPNMM!Transition( transitionArcs <- tAA, transitionArcs <- tAS,
      event <- 's.'.concat(lscAM.content),
      mpn <- thisModule.resolveTemp(lscAM.endLocation.object.lsc, 'mpn') ),
    tAA : MPNMM!TransitionArc( place <- sPA ),
    tAS : MPNMM!TransitionArc( place <- sPS ),
    pAA : MPNMM!PlaceArc( transition <- tA, place <- thisModule.getPlaceOfPrevMsgSender(lscAM) ),
    ...
}
rule bypassTransition{
  from
    firstLoc : LSCMM!Location,
    secondLoc : LSCMM!Location(firstLoc.isBypassCombination(secondLoc))
  to
    pA : MPNMM!PlaceArc( transition <- t,
      place <- secondLoc.getPlaceForLocAndInst ),
    t : MPNMM!Transition( event <- secondLoc.getEventName, transitionArcs <- tA,
      mpn <- thisModule.resolveTemp(firstLoc.object.lsc, 'mpn') ),
    tA : MPNMM!TransitionArc( place <- thisModule.getPlaceForLocAndInst(secondLoc) )
}
rule bypassTransitionWithFollowingAsyncSend extends bypassTransition{
  from
    firstLoc : LSCMM!Location,
    secondLoc : LSCMM!Location(firstLoc.isBypassCombination(secondLoc)
      and secondLoc.isAsyncSendingLocation)
  to
    tAS : MPNMM!TransitionArc( transition <- t,
      place <- thisModule.resolveTemp(secondLoc.outgoingMessage, 'sPS')
    ),
    pAS : MPNMM!PlaceArc ( place <- thisModule.resolveTemp(secondLoc.outgoingMessage, 'sPS'),
      transition <- thisModule.resolveTemp(
        Tuple{fL = firstLoc.getOppositeLocation, sL = secondLoc.getOppositeLocation}, 't') )
}
```

Fig. 4. SDM and ATL rules for asynchronous messages (T3)

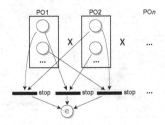

Fig. 5. Synchronization of open places at the end of an MPN (T4)

In ATL the transformation, depicted in Figure 4, is split into three rules that match different source elements. Where *AsyncMPNPattern* matches all messages in the *LSC* model and translates the messages, the two *bypassTransition* rules create the additional transitions for cold messages. For that, the messages have to be matched again. This can be done by splitting the transformation in two sequential transformations, losing the traceability links of the former steps, or as done here, by matching other source elements (combination of locations). The *to* parts of the ATL rules and their helpers extensively use traceability links.

(T4) The LSC presented in Section 2 ends with a hot message, so that only the places belonging to the last message have to be synchronized via a transition with `stop` event. For handling one or more cold messages at the end of a chart, all combinations of places that have not been synchronized from different LSC objects have to be connected to an own transition. As shown in Figure 5, the Cartesian product $(po1, po2, \ldots) \in PO1 \times PO2 \times \ldots$ has to be derived, where POn contains all unsynchronized places of the n-th *LSCObject*.

Such a pattern with two variable dimensions – the number of *LSCObjects* and the number of *Places* for each object – cannot be described in a declarative manner in ATL and SDM. Therefore, a recursive traversal is needed that can only be realized imperatively. In SDM, this can be handled locally in the control flow by calling rules recursively, whereas, in ATL this has to be implemented fully imperatively in global helpers.

4.2   Evaluation of the Transformations

To validate the semantical equivalence of the two transformations, both approaches have been fed with the same set of input models, and the output models have been compared manually with each other. The input models with up to five messages have been chosen to cover different sequences of messages based on the temperature, the type, and the direction between the two LSC objects. Caused by the manual review of the output models, these input models are as compact as possible to realize all combinations with respect to the specifications of the transformations.

To evaluate the equivalence for more complex models, an "automatic" comparison[2] of the output MPNs is needed. Therefore, the part of the SDM transformation that is presented in this paper has been transferred to the eMoflon

---

[2] Test suite provided at: http://www.moflon.org/emoflon

tool, a new version of MOFLON based on EMF. Hereby, EMFCompare allows for an automatic comparison of the output models (MPNs).

5   Desirable Features

In this section, proposals for additional features of these two languages are made. These originate from *(R)* the basic requirements of Section 3, *(T)* the transformation in Section 4, and *(S)* additional requirements for the monitor generation process scenario.

Implicit Traceability Links (R3, T2, T3). The explicit modeling of traceability links in SDM alloy the readability of the transformation rules caused by the additional patterns to create these links. An implicit generation as realized for subgraph copying [19] would be desirable for the here studied more general case.

In-Place Transformation (R5, R6). For post-processing purposes in the MPN target model SDM, as an in-place language, is favorable, because recursive deletions and modifications can be modeled within the control flow. Using the refining mode of ATL 3.1 (realized by copy rules) allows a kind of in-place transformation, but this approach fails when post-processing steps have to be repeated iteratively on the changed model until there is no new match. This is caused by the write-only target model that has to be used as source model in a repeated external call of the transformation. In the implementation of ATL 3.2 an extended support for in-place transformations and explicit deletions has been added, but with the drawback that some advanced imperative features are not supported.

Patterns of Dynamic Size (R7, T4). In complex transformations some problems occur that are typically resolved in programming languages with a recursive approach, which is needed to compute the Cartesian product of an unknown number of sets each containing an unknown number of elements. Figure 5 shows this using the example of the synchronization of all places at the end of the transformation, presented in Section 4.1. To eliminate the imperative part for this issue, a template concept for patterns is needed that allows the dynamic runtime initialization of parts of patterns by a quantity of instances in the model.

Explicit Modeling of Control Flow (T1, T3). When a complex transformation should be described in ATL, every object can only be bound by one rule, which leads to a shortage of unbound elements for other rules, as presented in Section 4.1 in *T3*. This forces the developer to design more complex holistic rules or split the complete transformation into independent sequential steps. By splitting the transformation, traceability links created in a previous step are not accessible in the current step.

Reusability of Matched Patterns (T2). In SDM, set patterns can be used to match many objects of the same type at once. The results can be passed between rules and returned as result but cannot be used for further pattern matching. By extending the set concept and allowing, additionally, the passing of matched patterns, the control flows in SDM rules could be reduced.

Deterministic and Correct Result (S). In the presented monitor generation process the correctness of the resulting model has to be ensured. Therefore,
properties such as a deterministic generation of target models, as guaranteed
by the declarative part of ATL [8], are desirable. As shown in the previous section, it is often impossible to provide a purely declarative solution for complex
transformations. Hence, in both ATL and SDM the developer has to cope with
non-determinism in the modeled transformation. To address this issue, test practices, as suggested in [3], have to be developed for ATL and SDM.

Integration into Software (S). The tools for ATL and SDM provide different approaches for the integration of transformations in software products.
MOFLON and FUJABA use SDM specifications to generate Java code and
ATL is translated into byte-code that is interpreted by a special virtual machine (ATL VM). Hence, the SDM code is preferable for seamless integration
in a standalone tool [6]. When developing a tool integrated in Eclipse, both
approaches are suitable.

6   Conclusion and Future Work

In this paper, we have presented a case study about the comparison of the
transformation languages ATL and SDM in the context of a model-based security monitor development process. We have highlighted shortcomings that have
evolved during the case study and suggested additional concepts to improve the
modeling of transformations. Both languages lack some features and should be
extended. One major disadvantage of ATL is the missing possibility to explicitly
model the control flow, and the resulting problem that elements can be bound
only once in a transformation.

A more satisfactory model transformation language for our monitor generation
process could be based on an SDM-like hybrid language that is extended by
static type and determinism analysis from PROGRES and Critical Pair Analysis
from AGG [12]. The language should support recursive patterns as implemented
in VIATRA. Additionally, a possibility for implicit traceability links should be
supported. Furthermore, an improved parameter handling for passing matched
patterns and set patterns between part rules is also desirable.

As stated, all these concepts have been proposed for different transformation
languages, but were never combined in an implementation of a graph transformation language. Therefore, further research has to determine the compatibility
of these extensions, e.g., determinism and recursive patters.

References

1. Amelunxen, C., Königs, A., Rötschke, T., Schürr, A.: MOFLON: A Standard-
   Compliant Metamodeling Framework with Graph Transformations. In: Rensink,
   A., Warmer, J. (eds.) ECMDA-FA 2006. LNCS, vol. 4066, pp. 361–375. Springer,
   Heidelberg (2006)
2. van Amstel, M., Bosems, S., Kurtev, I., Ferreira Pires, L.: Performance in Model
   Transformations: Experiments with ATL and QVT. In: Cabot, J., Visser, E. (eds.)
   ICMT 2011. LNCS, vol. 6707, pp. 198–212. Springer, Heidelberg (2011)

3. Baudry, B., Dinh-Trong, T., Mottu, J.M., Simmonds, D., France, R., Ghosh, S., Fleurey, F., Le Traon, Y.: Model transformation testing challenges. In: ECMDA Workshop on Integration of MDD and MDT. IRB Verlag (2006)
4. Czarnecki, K., Helsen, S.: Feature-based survey of model transformation approaches. IBM Systems Journal 45, 621–645 (2006)
5. Di Ruscio, D., Jouault, F., Kurtev, I., Bézivin, J., Pierantonio, A.: Extending AMMA for supporting dynamic semantics specifications of DSLs. Tech. rep. LINA (2006)
6. Fischer, T., Niere, J., Torunski, L., Zündorf, A.: Story Diagrams: A New Graph Rewrite Language Based on the Unified Modeling Language and Java. In: Ehrig, H., Engels, G., Kreowski, H.-J., Rozenberg, G. (eds.) TAGT 1998. LNCS, vol. 1764, pp. 296–309. Springer, Heidelberg (2000)
7. Hildebrandt, S., Wätzoldt, S., Giese, H.: Executing graph transformations with the MDELab story diagram interpreter. In: Transformation Tool Contest (2011)
8. Jouault, F., Kurtev, I.: Transforming Models with ATL. In: Bruel, J.-M. (ed.) MoDELS 2005. LNCS, vol. 3844, pp. 128–138. Springer, Heidelberg (2006)
9. Jouault, F., Allilaire, F., Bézivin, J., Kurtev, I.: ATL: A model transformation tool. Science of Computer Programming 72(1-2), 31–39 (2008)
10. Klar, F., Rose, S., Schürr, A.: TiE – a tool integration environment. In: Proc. of the 5th ECMDA-TW. CTIT Workshop Proc., vol. WP09-09, pp. 39–48 (2009)
11. Kolovos, D.S., Paige, R.F., Polack, F.A.C.: The Epsilon Transformation Language. In: Vallecillo, A., Gray, J., Pierantonio, A. (eds.) ICMT 2008. LNCS, vol. 5063, pp. 46–60. Springer, Heidelberg (2008)
12. Mens, T., Taentzer, G., Runge, O.: Detecting structural refactoring conflicts using critical pair analysis. In: Proc. of the Workshop on Software Evolution through Transformations. ENTCS, vol. 127, pp. 113–128. Elsevier (2005)
13. Meyers, B., Van Gorp, P.: Towards a hybrid transformation language: Implicit and explicit rule scheduling in story diagrams. In: Proc. of the 6th Int. Fujaba Days (2008)
14. OMG: MOF 2.0 QVT Spec. Object Management Group (January 2011), http://www.omg.org/spec/QVT/1.1/
15. Patzina, S., Patzina, L., Schürr, A.: Extending LSCs for Behavioral Signature Modeling. In: Camenisch, J., Fischer-Hübner, S., Murayama, Y., Portmann, A., Rieder, C. (eds.) SEC 2011. IFIP AICT, vol. 354, pp. 293–304. Springer, Heidelberg (2011)
16. Schürr, A.: Programmed Graph Replacement Systems. In: Handbook of Graph Grammars and Computing by Graph Transformation, vol. 1: Foundations, pp. 479–546. World Scientific (1997)
17. Taentzer, G., Ehrig, K., Guerra, E., de Lara, J., Lengyel, L., Levendovszky, T., Prange, U., Varró, D.: Varró-Gyapay, Sz.: Model transformation by graph transformation: A comparative study. In: Proc. of Workshop MTiP (2005)
18. Troya, J., Vallecillo, A.: Towards a Rewriting Logic Semantics for ATL. In: Tratt, L., Gogolla, M. (eds.) ICMT 2010. LNCS, vol. 6142, pp. 230–244. Springer, Heidelberg (2010)
19. Van Gorp, P., Schippers, H., Janssens, D.: Copying subgraphs within model repositories. In: Proc. of the 5th Int. Workshop on Graph Transformation and Visual Modeling Techniques. ENTCS, vol. 211, pp. 133–145. Elsevier (2008)
20. Varró, D., Balogh, A.: The model transformation language of the VIATRA2 framework. Science of Computer Programming 68(3), 214–234 (2007)
21. Wimmer, M., Kappel, G., Kusel, A., Retschitzegger, W., Schönböck, J., Schwinger, W., Kolovos, D., Paige, R., Lauder, M., Schürr, A., Wagelaar, D.: A Comparison of Rule Inheritance in Model-to-Model Transformation Languages. In: Cabot, J., Visser, E. (eds.) ICMT 2011. LNCS, vol. 6707, pp. 31–46. Springer, Heidelberg (2011)

Applying Advanced TGG Concepts for a Complex Transformation of Sequence Diagram Specifications to Timed Game Automata[*]

Joel Greenyer[1,**] and Jan Rieke[2,***]

[1] Politecnico di Milano, Piazza Leonardo Da Vinci, 32, 20233 Milano, Italy
greenyer@elet.polimi.it
[2] University of Paderborn, Zukunftsmeile 1, 33102 Paderborn, Germany
jrieke@uni-paderborn.de

Abstract. Declarative model transformation languages like QVT-R and TGGs are particularly convenient because mappings between models can be specified in a rule-based way, describing how patterns in one model correspond to patterns in another. The same mapping specification can be used for different transformation and synchronization scenarios, which are important in model-based software engineering. However, even though these languages already exist for a while, they are not widely used in practice today. One reason for that is that these languages often do not provide sufficiently rich features to cope with many problems that occur in practice. We report on a complex model transformation that we have solved by TGGs. We present advanced extensions of the TGG language that we have integrated in our tool, the TGG INTERPRETER.

Keywords: model transformation, Triple Graph Grammar (TGG), case.

1 Introduction

Declarative model transformation languages like QVT-Relations and TGGs are particularly convenient because mappings between models can be specified in a rule-based way, describing how particular patterns in one model correspond to particular patterns in another. The same mapping specification can often be interpreted for different application scenarios, e.g., for the forward transformation from a given source model to a target model or for the backward transformation from a given target model to a source model. It can furthermore be used to keep corresponding models synchronized when changes occur to either one.

[*] This work was developed in the course of the Collaborative Research Center 614, Self-optimizing Concepts and Structures in Mechanical Engineering, Univ. of Paderborn, and was published on its behalf, funded by the Deutsche Forschungsgemeinschaft.
[**] This work was elaborated mainly while the author was working at the University of Paderborn, Germany.
[***] Supported by the International Graduate School Dynamic Intelligent Systems.

A. Schürr, D. Varró, and G. Varró (Eds.): AGTIVE 2011, LNCS 7233, pp. 222–237, 2012.

However, even though these languages already exist for a while and a range of (mostly academic) tools have been developed in the past, these languages are not widely used in practice today. One reason for that is that these languages often do not provide sufficiently rich features to cope with many practical transformation problems. As a consequence, the formalisms may seem appealing at first, but many developers faced with real-life problems quickly return to "program" their transformations, using an operational language.

In this paper, we report on a complex model transformation that we have solved by TGGs (Sect. 3). We present advanced extensions of the TGG language that we have integrated in our tool, the TGG INTERPRETER. First, we describe the integration of OCL for describing attribute value constraints and application conditions (Sect. 4). We especially support the definition of custom operations that can be reused in the TGG rules, making them more readable.

Second, we present how constraints on stereotypes in UML domains can be conveniently specified in the TGG rules (Sect. 5). This extension is crucial because today many specific languages are defined by providing profiles for UML.

Third, we present a rule generalization concept, revising the one presented earlier by Klar et al. [15] (Sect. 6). By using generalization, we greatly reduced the number of redundant patterns that needed to be specified for our example.

Last, in our case study we experienced that there are many transformation rules where in some cases we wish to create elements in the target model, but in other cases we wish to reuse elements or whole patterns that were created in the target model by previous rule applications. We present an advanced concept for controlling the reuse of model patterns in the target domain in Sect. 7.

Furthermore, we informally discuss important properties of our TGG extensions and the transformation algorithm in Sect. 8, report on related work in Sect. 9, and conclude in Sect. 10. But first, we briefly introduce TGGs.

2   Triple Graph Grammars

Triple Graph Grammars (TGGs) [20] allow us to define sets of corresponding graphs. An element of this set is typically a triple consisting of two independent graphs that are linked via a third graph, called the *correspondence* graph. Because of this triple structure, such a graph is also called a *triple-graph*. These different graphs in a triple-graph are typed over different type graphs. TGG rules are non-deleting graph production rules that describe how, based on a start graph or *axiom*, triple-graphs can be created. Triple-graphs that can be created by a TGG are called *valid* triple-graphs.

Transferred to the "modeling world", TGGs define sets of corresponding models, also called *triple-models*, where the independent models, called *domain models*, are instances of different meta-models. The domain models are linked via a *correspondence model*, which is an instance of a correspondence meta-model.

TGGs can be interpreted for different *application scenarios*. In this paper, we focus on the *forward transformation* scenario: A model of one domain is given, called the *source* domain in the following. A TGG can now be operationalized

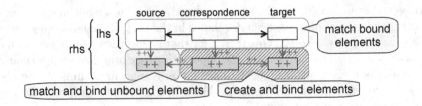

Fig. 1. The interpretation of a TGG rule for the forward transformation

to create a model of the opposite domain, called the *target* domain, and a correspondence model, such that the resulting models form a valid triple-model.

A TGG rule consists of nodes and edges that represent objects and links in the domain models. Since a TGG rule is a non-deleting graph grammar rule, the nodes and edges appear either on the left-hand side (lhs) *and* right-hand side (rhs) of the rule, or they appear on the right-hand side *only*. The former nodes and edges are also called *context nodes* and *context edges*, the latter are called *produced nodes* and *produced edges*. Context nodes are displayed as white boxes with a black border; produced nodes are displayed as green boxes with a dark green border and a "++" label. Context edges are displayed as black arrows; produced edges are displayed as dark green arrows with a "++" label.

Fig. 1 shows an abstract illustration of a TGG rule and how it is *applied* in a forward transformation scenario. Consider a state during the transformation where some rules were already applied and some elements in the source model were already translated to some target model elements. After a rule application, when an object in the model is matched by a node or created according to a node, we say that this object is *bound* to that node. A TGG rule is applied as follows: First, a match of the rule's context and source domain graph pattern must be found in the source model and the already created target and correspondence model. In this match, context nodes must be matched only to already bound objects and source produced nodes must only be matched to yet unbound objects. If such a match can be found, target and correspondence model elements can be created according to the produced target and correspondence pattern of the rule. All matched and created objects are bound to the according rule nodes. As a consequence, a model object can only be bound once to a produced node. We call this the *bind-only-once* semantics of produced nodes. Each model object in a valid triple-model is produced by exactly one produced TGG node of one TGG rule application. Thus, the bind-only-once semantics ensures that the resulting models form a valid triple-model according to the TGG.[1]

In our TGG INTERPRETER, we not only track which objects are bound to which nodes. We also track which links are bound to which edges. The set of all node and edge bindings after a rule application is called a *rule binding*.

---

[1] Most TGG transformation engines, like MOFLON [1], essentially implement the same semantics. However, these tools mostly do not capture a node/object binding in an explicit data structure: an object is considered bound if there is a link from a correspondence object pointing to it.

Also constraints on attribute values and application conditions can be formulated in a TGG rule, as explained in more detail in Sect. 4. Furthermore, we have introduced the concept of *reusable nodes* and *reusable edges*, displayed in gray with a "##" label. They can be interpreted either as produced nodes and edges or as context nodes and edges [10], as explained in more detail in Sect. 7.

3   Example

The transformation example is a mapping from *Modal Sequence Diagram* (MSD) specifications to networks of *Timed Game Automata* (TGA), which is performed in order to find inconsistencies in the specification with UPPAAL TIGA [3,9].

MSDs: MSDs are a formalism for specifying interactions among objects that may, must, or must not happen, proposed as a formal interpretation of UML sequence diagrams by Harel and Maoz [13]. In an MSD specification, the interaction among system and environment objects is specified in sequence diagrams where messages have a *hot* or *cold* temperature. The left of Fig. 2 shows an MSD specification with three MSDs. Hot messages are displayed as solid red arrows; cold messages are displayed as dashed blue arrows. On the top left, a collaboration diagram shows the object system, which here consists only of an environment object e:Env and a system object s:S.

In short, the semantics of an MSD specification is as follows: If in a sequence of interactions a message is sent in the system that corresponds to the first message in an MSD, an *active copy* of that MSD, or *active MSD*, is created. (We only allow a single active copy of an MSD at a time.) Upon the occurrence of further messages in the system that correspond to the subsequent messages in the MSD, the active MSD progresses. This progress is captured by the *cut*, which marks the occurred messages in the active MSD. If the cut is immediately prior to a message on its sending and receiving lifeline, this message is *enabled*. If a hot and executed message is enabled, it means that this message must eventually occur and that no message must occur that corresponds to another message in the diagram that is not currently enabled. Due to these liveness and safety requirements, there can be *inconsistencies* in an MSD specification. An MSD specification is *inconsistent* if there exists a sequence of environment events for which the system objects cannot avoid a violation of these requirements.

Timed Game Automata in Uppaal Tiga: UPPAAL TIGA is an extension of the UPPAAL model checker [2]. In UPPAAL, a system is modeled as a network of Timed Automata (TA). Such a TA network consists of parallel automata that each consist of *locations* and *edges*. The edges in the parallel automata can be synchronized via *channels*. A *transition* in a TA network is one edge or multiple synchronized edges firing in the parallel automata. The edges can also have guard conditions and update expressions that assign values to variables. Variables and side-effect-free functions can be declared *globally* for the whole TGA network or *locally*, only visible within one automaton in the network.

In UPPAAL TIGA, the Timed Automata are extended to Timed Game Automata (TGA), in which the edges can be either *controllable* or *uncontrollable*.

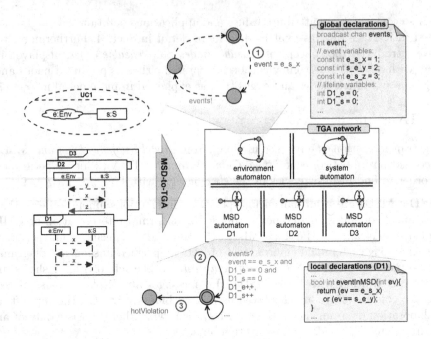

Fig. 2. MSD-to-TGA transformation overview

If only controllable edges participate in a transition, the transition is controllable by the *system*; if at least one edge is uncontrollable, the transition is controllable by the *environment*. UPPAAL TIGA can check different kinds of properties in a TGA network, for example if some state is reachable by the system even though the environment will always try to keep the system from reaching that state [3].

The MSD-to-TGA Mapping: An MSD specification can be mapped to a TGA network so that UPPAAL TIGA can check whether the system is always able to avoid a state that corresponds to a violation of the requirements [8,9]. If that is the case, the MSD specification is consistent, otherwise it is inconsistent.

The mapping principle is illustrated in Fig. 2: For an MSD specification, one *environment automaton* and one *system automaton* is created. For each MSD in the specification, one *MSD automaton* is created. Together, these automata form a TGA network. The environment and system automata encode the behavior of the environment and system objects sending messages in the system. The MSD automata encode the progress of the cut in the active MSD and violations that may occur in the MSD. The cut is encoded by globally declared *lifeline variables* that are created for each lifeline in each MSD.

If the environment chooses to send message x from object e:Env to the object s:S, this works as follows in the TGA network. First, the environment takes an edge in the environment automaton, assigning an according constant value to the variable event (①). Then the environment automaton takes an edge that emits over the broadcast channel events. This may synchronize edges in the MSD automata that represent the message. For each message in the MSD there is an edge in the MSD automaton. This edge has a guard that ensures that it is only synchronized if the message sent is enabled in the current cut of the active MSD. It has an update label where the lifeline variables corresponding to the message's sending and receiving lifelines are increased, encoding the progress of the cut. Fig. 2 shows the edge (②) that corresponds to the first message in MSD D1.

Each message in each MSD is furthermore mapped to an integer constant declaration that represents the message in the above process. The constant name for a message has the form <name of sending object>_<name of receiving object>_<name of message>. The constant value is always the value of a previously created constant plus one. Of course, there must not be two variables or constants with the same name in the TGA specification. Thus, many diagram messages may be mapped to the same constant declaration if they have the same sender, receiver and message name. For example, the three messages called x in the MSDs D1 and D2 must all be mapped to a single declaration of the constant e_s_x. Similarly, each message in each MSD is mapped to an edge in the environment or system automaton (depending on whether it is a message sent by an environment or system object), which assigns the corresponding constant to the variable event (Fig. 2 (①)). Again, there must not be two edges that assign the same value to the event variable in the environment or MSD automaton.

For each MSD, furthermore an edge is created in the MSD automaton that is taken if a message is sent that is violating the active MSD in the current cut, i.e., the according message is not currently enabled, but nevertheless appears in the MSD. The guard and update labels of this edge (Fig. 2 (③)) are not shown in detail here. What's more important is that, in order to know whether a currently "sent" message appears in the MSD, a Boolean function eventInMSD(int ev) is created in the local declarations of each MSD automaton for each MSD in the specification. This function has a return statement that consists of a disjunction that for each message in the MSD contains a statement that renders the disjunction true if the value of the variable event corresponds to that message in the MSD. There must not be two redundant sub-expressions in the disjunction, so there is for example only one sub-expression (ev == e_s_x) even though the message x appears in the MSD D1 two times.

In summary, the transformation from MSD to TGA is complex for the following reasons. First, the resulting TGA models are complex, and, second, we have to distinguish several different cases when translating elements (e.g., different kinds of messages: hot and cold; messages sent from environment or system objects; messages at the beginning, middle, or end of an MSD). Third, complex string concatenations are required for variable and constant definitions, and, fourth, certain elements must not exist twice in the target model.

We realized this mapping by a TGG transformation from UML to an EMF[2]-based UPPAAL TIGA model. The transformation can be downloaded as part of SCENARIOTOOLS.[3] The TGG INTERPRETER can also be installed separately.[4]

---

[2] http://www.eclipse.org/emf/

[3] http://www.cs.uni-paderborn.de/index.php?id=scenariotools

[4] http://www.cs.uni-paderborn.de/index.php?id=tgg-interpreter

228   J. Greenyer and J. Rieke

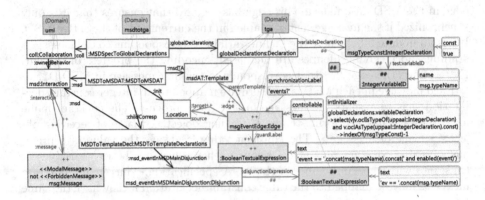

Fig. 3. Message2Edge: A TGG rule for translating messages

4 OCL Integration

This transformation is an example where many string attribute values in the
target model must be concatenated from different pieces of information in the
source model. In order to describe such string concatenations and other queries
on the models, we have integrated OCL [18] in TGGs.

In a TGG rule, OCL expressions can be used in *attribute value constraints*
and *application conditions*. They are displayed as yellow rounded rectangles in
the TGG rule. In the OCL expressions, the names of nodes in the same rule can
be used as variables, which are bound to the same object as the node is bound
to when the rule is matched in the model (or when target model objects are
created). The TGG rule in Fig. 3 shows a range of attribute value constraints.

This rule maps messages in the UML model to a range of elements in the TGA
model. This diagram shows the concrete syntax of the TGG rule editor that is
part of our TGG INTERPRETER tool suite. The domains are represented by the
violet nodes at the top of the diagram that link the nodes in the domain via the
thin dotted gray lines. In this rule, a message in the UML model is mapped to
an edge in the MSD automaton, represented by the node msgEventEdge:Edge.
This edge is added to the automaton represented by the node msdAT:Template
(an automaton definition is called a *template* in UPPAAL). The source and target
location of the edge is the location represented by the node :Location. Furthermore, the rule maps a message to the corresponding global constant declaration
and the sub-expression of the return statement in the function eventInMSD(int
ev) that is declared locally for the corresponding MSD automaton (see bottom
right in Fig. 2).

An attribute value constraint always points to a node, which is called its
slot node. The top row of the attribute value constraint's rounded rectangle
specifies the constrained attribute, which is called the constraint's *slot attribute*.
The bottom row specifies an OCL expression, which is called the constraint's
value expression. An attribute value constraint specifies that the slot node can

only be bound to an object if the value of the slot attribute equals the value specified by the value expression. During a forward transformation, attribute value constraints in the target domain are interpreted as assignments.

Application conditions are also displayed as yellow rounded rectangles, but they do not specify a slot node or slot attribute. They only specify an OCL expression, called the *condition expression*, which must evaluate to a Boolean value. Application conditions come in two flavors: They can be either *preconditions* or *postconditions*. Preconditions must evaluate to true in order to apply the rule. At the end of the transformation, all postconditions of all applied rules must evaluate to true.

Operationally, the attribute value constraints are considered as follows. We consider a forward transformation scenario for simplicity. The TGG engine employs a graph matching algorithm that starts with an initial matching of some source or context node in the TGG rule, and tries to find a pattern in the domain models that is isomorphic to the source and context pattern (as explained in Sect. 3). During the graph matching process, the TGG engine tries to evaluate the value expression of an attribute value constraint as soon as a candidate object for matching the slot node is found. If the result of the evaluation equals the value that the candidate object carries for the slot attribute, we say that the attribute value constraint *holds*. A node can be bound to an object if all attached attribute value constraints hold; then the graph matching can continue, otherwise the algorithm backtracks.

When evaluating the value expression, it may however happen that a variable in the expression is unbound because it corresponds to a node in the rule that is not yet bound. In this case, the attribute value constraint is marked for a delayed evaluation. It is evaluated as soon as all the nodes that appear as variables in its value expression are bound. If the constraint holds, the graph matching continues. If it turns out that the constraint does not hold, the graph matching backtracks. When backtracking, another binding for nodes that appear as variables in the constraint's value expression may be found so that the attribute value constraint holds, but the graph matching may also backtrack to find another candidate object for the constraint's slot node.

When creating elements in the target model, the attribute value constraints are interpreted as assignments. This means that when an object is created in the target model according to a node, the value expression of each attached attribute value constraint is evaluated. The value is then assigned to the slot attribute of the object created for the slot node. There may also be a delayed evaluation if the value expression of one constraint refers to a node that was not yet created.

At the end of the transformation, all attribute value constraints of all applied TGG rules are checked once again. The transformation is only correct if all attribute value constraints hold.

Preconditions are evaluated as soon as all the nodes that appear as variables in the condition expression are bound. The graph matching backtracks if the condition expression evaluates to false. Postconditions are evaluated for all rule applications at the end of the transformation. The transformation is only correct if all postconditions are satisfied.

The above section for example mentions the naming scheme for the constant that represents a particular message in the TGA network. This name appears not only in the global declarations, but also in the update and guard labels and the local declarations of the different automata. In order to avoid that complex OCL expressions occur redundantly in the TGG rules, the TGG INTERPRETER allows the transformation engineer to define custom derived attributes for domain model elements within the transformation definition. These custom attributes can be defined in a separate OCL file and the OCL expressions within the TGG rules can refer to these attributes. For the MSD-to-TGA mapping, we have for example defined the custom derived attribute typeName for UML messages. It produces the string <name of sending object>_<name of receiving object>_<name of message> as explained above. This derived attribute is used three times within OCL expressions in the TGG rule shown in Fig. 3.

5   Stereotype Constraints

With the powerful UML tools that are being developed around the Eclipse implementation of UML2,[5] model-based development in practice increasingly employs UML and its lightweight profile extension mechanism [19]. We have for example used a profile to add the temperature attribute of messages in MSDs to UML sequence diagrams (similar to Harel and Maoz [12]) or to mark objects in the collaboration diagram as system or environment objects (see Fig. 2).

In transformations involving stereotyped UML models, it is crucial to be able to specify that a certain UML object has a particular stereotype applied or not. For this purpose, we have extended TGG rules by *stereotype constraints* that can be added to nodes in UML domains. Stereotype constraints specify that a node can only bind a UML object if a certain stereotype is or is not applied to that element. These constraints are shown within a node's label. An entry in double angle brackets represents a required stereotype. If preceded by the keyword not, this stereotype must not be applied.

The node msg:Message of the rule in Fig. 3 shows an example where the stereotype ModalMessage must be applied and the stereotype ForbiddenMessage must not be applied. When a UML object is created according to a node with a positive stereotype constraint, the stereotype is applied to the object.

If a positive stereotype constraint is added to a node, it is possible to add attribute value constraints where the slot attribute is an attribute defined by the stereotype that the stereotype constraint refers to. The TGG rule in Fig. 4 shows an example where the attribute value constraint added to the node spec-Part:Property refers to the attribute partKind, an attribute defined by the stereotype SpecificationPart.

---

[5] http://www.eclipse.org/uml2/

6   TGG Rule Generalization

Generalization is a powerful mechanism in object-orientation for reusing and extending existing solutions. Klar et al. have introduced this concept to TGG rules [15] and realized it within the MOFLON tool suite. In our example transformation, there are different kinds of messages that need to be mapped to the TGA model. Some elements in the TGA model must be created for all messages, some must be created e.g., only for environment messages. For this purpose, we adopted the rule generalization concept proposed by Klar et al. with some improvements.

A TGG rule describes a relation between sets of objects. Klar et al. argue that "generalization usually means that a member of a more specialized type also is a member of the more general type" and, thus, whenever a more specialized TGG rule is applicable, also the more general rule should be applicable [15, Sect. 4.1]. We call this the *guiding principle* of TGG rule generalization in the following.

To ensure that, Klar et al. define a number of syntactical constraints for a TGG rule that specializes another. These constraints require, first (1), that a specialized rule contains a copy of the general rule [15, rule 14]. Second (2), context nodes in the specialized rule may be replaced by nodes with a more specialized type [15, rule 15]. Third (3), produced nodes in the general rule can be converted to context nodes in the specialized rule [15, rule 15]. Forth (4), new nodes and edges may be added to the context and produced pattern of the rule [15, rule 16], and, fifth (5), further attribute value constraints and application conditions may be added to the rule [15, rule 17]. Last (6), the specialized rule must have a higher *priority* than the more general rule [15, rule 10]. Priorities are numbers assigned to rules; the MOFLON transformation engine will first try to apply rules with a higher number. This way it is ensured that a more general rule is only applied when any specialization of that rule cannot be applied.

In the TGG INTERPRETER, a specialized TGG rule also basically consists of a copy of the more general rule (as (1) above) and nodes, edges, and constraints may be added to the specialized rule (as (4) and (5) above). Furthermore, nodes may be replaced by nodes with a more special type class. In contrast to (2) above, this is *allowed* also for produced nodes, since it does not violate the guiding principle of TGG rule generalization. However, it is *not allowed* to convert produced nodes to context nodes in the more specialized rule (as (3) above). This is not allowed because, due to the bind-only-once semantics of produced nodes, this would violate the guiding principle of TGG rule generalization.

Another difference in the TGG rules of the TGG INTERPRETER tool suite compared to the MOFLON tool suite is that in order to create a specialized rule, the transformation engineer does not literally need to create a copy of the more general rule first. Instead, the rule diagram of the more specialized rule just contains the added nodes, edges, and constraints, and such nodes from the more general rule to which additional edges and constraints are attached. Also, the specialized rule contains the nodes from the more general rule which are given a more specialized type. All other nodes from the more general rule are only "copied" into the specialized rule during transformation-time. This makes the rule set better maintainable and the rule diagrams more concise.

232   J. Greenyer and J. Rieke

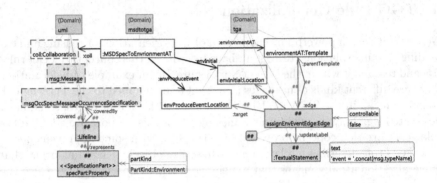

Fig. 4. EnvironmentMessage2Edge: A TGG rule for translating a minimal environment message

Nodes of a more general rule that recur in the specialized rule are called *refining nodes*. They have the same name as the node that they represent and are displayed with a dashed border. Figure 4 shows a specialization of the TGG rule in Fig. 3. This specialized rule maps an environment message (also) to an edge in the environment automaton. The nodes coll:Collaboration and msg:Message are refining nodes that appear in this rule because patterns that are added in this specialized rule are attached to these nodes. A message is an environment message if the SpecificationPart stereotype is applied to the property in the collaboration diagram that is represented by the sending lifeline of the message; moreover, the stereotype application must carry the value Environment for the partKind attribute. This is expressed by the pattern added to the UML domain of this rule. In the target domain, the message is mapped to an edge in the environment automaton with an update label as explained in Sect. 3.

Another difference to the rule generalization approach presented by Klar et al. is that we do not use priorities to ensure that the transformation engine will always try to apply more specialized rules before trying to apply more general rules. Instead of priorities we define that a more specialized rule has *precedence* over its more generalized rule. The transformation engine will not try to apply a rule if it did not try to apply another rule with precedence over that rule. The difference in this approach is that the precedence induces a *partial order* among the TGG rules whereas the priorities define a *total order*. The advantage of the precedence is that it is less restrictive and will not unnecessarily restrict the non-determinism among the rules; if we for example employ heuristics for applying TGG rules in a smart order for increasing the transformation speed, such heuristics will have more freedom to select the next rule to apply.

Note that the precedences are only relevant in the operational interpretation of the rules, i.e., they are only a directive for the transformation engine to try to apply certain rules first in a particular application scenario. By contrast, the valid triple-models are defined as those that can be produced by the TGG rules regardless of the precedences.

7   Reusable Patterns

As described in Sect. 3, each diagram message in each MSD is mapped to an integer constant declaration. The "same" diagram message can appear several times in several MSDs and must be mapped to the same constant declaration. To handle the case where yet no constant declaration exists for a message, we would need a rule where this constant declaration is represented by produced nodes. To handle the case where the constant declaration for a message already exists, we would need a rule where the constant declaration is represented by context nodes. The previous rule cannot be used for this case because of the bind-only-once semantics of the produced nodes. If there are many different elements that may or may not already exist, this leads to a large number of rules that must be created for mapping the same element. For that reason, we have introduced the concept of *reusable nodes* and *reusable edges* to TGGs [10]. The semantics of a rule with a reusable node is equivalent to two rules where the node is a produced node in one rule and a context node in the other. A transformation engine may therefore nondeterministically decide to interpret a reusable node as a produced node or as a context node. The same holds for reusable edges.

Reusable nodes can also appear in the source domain. The nodes representing the lifeline and the property in Fig. 4 are reusable nodes because they may or may not have been bound previously.

In the target domain, it is sometimes crucial to force the transformation engine to reuse a certain object structure, i.e., interpret the reusable nodes as context nodes. This is typically the case if creating another object structure instead of reusing one creates an invalid or inappropriate model. In the above case, there must for example not be two constant declarations with the same name. Furthermore, once an edge is created for an environment message in the environment automaton (see Fig. 4), this edge should be reused, because there should not be two edges from and to the same locations with identical guard, update and synchronization labels. Such constraints are sometimes part of a domain meta-model; sometimes they are only formulated for the purpose of a transformation. We call these constraints *global constraints* [10] and define that a triple-model produced by a TGG is only valid if all global constraints are satisfied. At the end of a transformation, our TGG INTERPRETER validates the constraints formulated in the domain meta-models and the transformation-specific global constraints that can be formulated via OCL in an external file.

If the global constraints are not satisfied at the end of a transformation, this means that the transformation engine may have to backtrack over the rule applications, reusing existing objects where previously it created them or creating new objects where it previously reused others. The latter could be required if global constraints formulate a lower bound, for example that there must be always at least two objects with certain properties in a model.

The TGG INTERPRETER, however, currently cannot backtrack over rule applications. Since in most cases the global constraints formulate upper bounds, such as that there must be only one object with certain properties in the model, it is in most cases sufficient to try reusing objects wherever this is possible.

The TGG INTERPRETER therefore implements a *reuse-before-create* semantics for reusable nodes. This is similar to the check-before-enforce semantics in QVT-Relations [17, Sect. 7.2.3]. In contrast to QVT-Relations, however, the reuse-before-create semantics is only *one possible operational interpretation* of reusable nodes in TGGs—it is *not* part of the general TGG semantics.

The reuse of an object or a link is decided for each reusable node and edge. Sometimes, however, this could lead to unintended effects. Consider the two reusable nodes assignEnvEventEdge:Edge and :TextualStatement in the TGG rule of Fig. 4. In a case where some environment messages were previously translated, there would be an uncontrollable edge in the environment automaton that the reusable node assignEnvEventEdge:Edge could always reuse. However, the update label statement attached to that edge may not be reusable, because the edge does not correspond to the "same" message. In this case, a second update label statement would be attached to the same edge, which is not what we intended.

As a solution, we have introduced the concept of *reusable patterns*. A reusable pattern is a set of reusable nodes and edges in a rule. It is represented by a small gray node with a "##" label that is connected to reusable nodes. The reusable pattern consists of all the connected nodes the reusable edges between them.

The semantics of a TGG rule with a reusable pattern is equivalent to two rules where all the nodes and edges in the pattern are produced nodes and edges in one rule and all the nodes and edges in the pattern are context nodes and edges in the other rule. Operationally, the TGG INTERPRETER will first try to reuse the pattern structure and will only try to create it if that is not possible.

8   Properties of the TGG Extensions

As described in Sect. 2 and 4, a triple-model is valid according to a TGG if (a) it can be produced by a sequence of TGG rule applications, (b) all postconditions and attribute value constraints hold for each applied TGG rule, and (c) all global constraint hold. If after a transformation all model domain elements are bound, the bind-only-once semantics and the final checking of above-mentioned constraints ensure that only valid triple-models are effectively created by the TGG INTERPRETER. This ensures the *correctness* of the results.

Also note that the precedence concept introduced for the rule generalization does not violate the correctness of a transformation. Intuitively, this is because the precedences are not considered when applying rules to produce the valid triple-models. Therefore, if a rule is applied in a forward transformation, the resulting bound triple-model always could have been created by a sequence of TGG rule applications that create all parts of the triple-model in parallel.

One general issue when operationally interpreting TGG rule in transformation scenarios is that at several steps during the transformation, different choices can be made on applying rules. This non-determinism leads to the problem that certain sequences of rule applications lead to producing a valid triple-model, but others do not. Our TGG INTERPRETER currently does not support backtracking over rule applications. Therefore, in some cases, we may not find a valid transformation result if one exists, i.e., our transformation algorithm is not *complete*.

Reusable nodes and rule inheritance potentially increase the variety of choices that the transformation engine has during a transformation. We plan to implement a backtracking mechanism in our TGG INTERPRETER. The backtracking mechanism should especially be able to consider choosing a more general rule instead of selecting only the most special ones applicable. It should also be able to change the interpretation for a reusable node or pattern (interpreting it as a produced node/pattern instead of a context node/pattern or vice versa).

As mentioned above, it may happen that several valid triple-models can be created from a source model, in which case the transformation result is not unique. We currently support no analysis methods that help to determine whether a transformation result is unique. Hermann et al. [14] present an approach that uses critical pair analysis to determine whether the transformation result of a TGG may not be unique.

9   Related Work

Dang and Gogolla [4] presented an approach for using OCL for specifying attribute value constraints and application conditions within TGGs. In their approach, they specify TGG rules textually, including a number of OCL statements. Then an OCL framework can execute the TGG rules, including the assignment of attribute values in the target domain. Compared to the approach presented here, however, they cannot define custom derived attributes for domain elements.

Golas et al. [7] show how to integrate application conditions in TGGs. They extend a formal framework for TGGs to show the termination, information preservation, correctness and completeness of transformations based on the extended TGGs. Their application conditions are restricted to formulating constraints on the context part of the rule. Also they assume that constraints in the source model are only evaluated in the scope of the already bound part of the source model. In the TGG INTERPRETER, constraints are evaluated with respect to the whole source model. We plan to investigate restrictions to our constraints that are necessary to ensure the completeness and information preservation of our transformations. Klar et al. [16] show that efficient translators for TGG with NACs are still preserving the fundamental TGG properties. However, these NACs are restricted to forbidding the existence of model elements.

To the best of our knowledge, there are no other TGG or QVT engines which provide a convenient support for constraints on stereotypes in UML models. Giese et al. [6] present a TGG-based transformation of a UML model with a SysML profile, but no indication is given on if and how constraints on stereotype applications are supported by their transformation engine.

Besides TGGs in MOFLON, we are not aware of another relational transformation engine supporting a rule generalization concept. The comparison with other, non-relational model transformation languages is beyond the scope of this paper. We refer to Wimmer et al. [21], who compare the TGG rule generalization concept of Klar et al. to the rule generalization concept of ATL[6] and ETL.[7] Guerra et al. [11] present a technique to specify transformations declaratively by relations between models that must or must not exist. Similar to rule generalization, it is a promising approach to make transformation specifications more intuitive. They also support attribute constraints.

Geiger et al. present a TGG engine [5] in which produced nodes can be matched multiple times to target objects. This violates the bind-only-once semantics for produced nodes, which is crucial for creating valid triple-models.

10   Conclusion and Outlook

In this paper, we reported on practically relevant TGG extensions that we elaborated and implemented in the TGG INTERPRETER. We extended TGGs by OCL for specifying attribute value constraints, application conditions, and custom attributes. We also integrated support for specifying constraints on stereotype applications and elaborated a rule generalization concept, refining the ideas of Klar et al. [15]. Last, we extended the concept of reusable nodes to reusable patterns to better control the reuse of model structures in target models.

With these extensions, TGGs become a powerful and flexible formalism for solving many complex model transformation and synchronization problems. We used these extensions in different transformation scenarios. Especially, the rule generalization improves the maintainability of the rule set. Complex OCL constraints and conditions also frequently occur in practical transformations.

We have also identified some open challenges. For example, using rule generalization in our example, we still ended up with some redundant rule patterns. We are therefore planning a more flexible rule extension mechanism. Furthermore, the reuse-before-create semantics of reusable patterns may not be practical in all cases. Therefore it could be useful to attach specific constraints on reusable patterns to more specifically control the reuse of model patterns. In addition, implementing backtracking over rule applications is also planned for the future.

References

1. Amelunxen, C., Königs, A., Rötschke, T., Schürr, A.: MOFLON: A Standard-Compliant Metamodeling Framework with Graph Transformations. In: Rensink, A., Warmer, J. (eds.) ECMDA-FA 2006. LNCS, vol. 4066, pp. 361–375. Springer, Heidelberg (2006)
2. Bengtsson, J., Larsen, K., Larsson, F., Pettersson, P., Yi, W.: UPPAAL – A Tool Suite for Automatic Verification of Real-time Systems. In: Alur, R., Henzinger, T.A., Sontag, E.D. (eds.) HS 1995. LNCS, vol. 1066, pp. 232–243. Springer, Heidelberg (1996)
3. Cassez, F., David, A., Fleury, E., Larsen, K.G., Lime, D.: Efficient On-the-Fly Algorithms for the Analysis of Timed Games. In: Abadi, M., de Alfaro, L. (eds.) CONCUR 2005. LNCS, vol. 3653, pp. 66–80. Springer, Heidelberg (2005)

---

[6] http://www.eclipse.org/atl

[7] http://www.eclipse.org/gmt/epsilon/doc/etl/

4. Dang, D.-H., Gogolla, M.: On Integrating OCL and Triple Graph Grammars. In: Chaudron, M.R.V. (ed.) MODELS 2008. LNCS, vol. 5421, pp. 124–137. Springer, Heidelberg (2009)
5. Geiger, N., Grusie, B., Koch, A., Zündorf, A.: Yet another TGG engine? In: Norbisrath, U., Jubeh, R. (eds.) Int. Fujaba Days. Kasseler Informatikschriften (2011)
6. Giese, H., Hildebrandt, S., Neumann, S.: Model Synchronization at Work: Keeping SysML and AUTOSAR Models Consistent. In: Engels, G., Lewerentz, C., Schäfer, W., Schürr, A., Westfechtel, B. (eds.) Graph Transformations and Model-Driven Engineering. LNCS, vol. 5765, pp. 555–579. Springer, Heidelberg (2010)
7. Golas, U., Ehrig, H., Hermann, F.: Formal Specification of Model Transformations by Triple Graph Grammars with Application Conditions. In: Rachid Echahed, A.H., Mosbah, M. (eds.) Int. Workshop on Graph Computation Models. Electronic Communications of the EASST, vol. 39 (2011)
8. Greenyer, J.: Synthesizing modal sequence diagram specifications with Uppaal-Tiga. Tech. Rep. tr-ri-10-310, University of Paderborn (2010)
9. Greenyer, J.: Scenario-Based Design of Mechatronic Systems. Ph.D. thesis, University of Paderborn (2011)
10. Greenyer, J., Kindler, E.: Comparing Relational Model Transformation Technologies: Implementing Query/View/Transformation with Triple Graph Grammars. Software and Systems Modeling 9(1), 21–46 (2010)
11. Guerra, E., de Lara, J., Orejas, F.: Pattern-Based Model-to-Model Transformation: Handling Attribute Conditions. In: Paige, R. (ed.) ICMT 2009. LNCS, vol. 5563, pp. 83–99. Springer, Heidelberg (2009)
12. Harel, D., Kleinbort, A., Maoz, S.: S2A: A Compiler for Multi-modal UML Sequence Diagrams. In: Dwyer, M.B., Lopes, A. (eds.) FASE 2007. LNCS, vol. 4422, pp. 121–124. Springer, Heidelberg (2007)
13. Harel, D., Maoz, S.: Assert and negate revisited: Modal semantics for UML sequence diagrams. Software and Systems Modeling 7(2), 237–252 (2008)
14. Hermann, F., Ehrig, H., Orejas, F., Golas, U.: Formal Analysis of Functional Behaviour for Model Transformations Based on Triple Graph Grammars. In: Ehrig, H., Rensink, A., Rozenberg, G., Schürr, A. (eds.) ICGT 2010. LNCS, vol. 6372, pp. 155–170. Springer, Heidelberg (2010)
15. Klar, F., Königs, A., Schürr, A.: Model transformation in the large. In: ESEC-FSE 2007, pp. 285–294. ACM, New York (2007)
16. Klar, F., Lauder, M., Königs, A., Schürr, A.: Extended Triple Graph Grammars with Efficient and Compatible Graph Translators. In: Engels, G., Lewerentz, C., Schäfer, W., Schürr, A., Westfechtel, B. (eds.) Graph Transformations and Model-Driven Engineering. LNCS, vol. 5765, pp. 141–174. Springer, Heidelberg (2010)
17. Object Management Group (OMG): MOF Query/View/Transformation (QVT) 1.1 Specification, OMG document formal/2011-01-01
18. Object Management Group (OMG): Object Constraint Language (OCL 2.2) specification, OMG document formal/2010-02-01
19. Object Management Group (OMG): UML 2.3 Superstructure Specification, OMG document formal/2010-05-03
20. Schürr, A.: Specification of Graph Translators with Triple Graph Grammars. In: Mayr, E.W., Schmidt, G., Tinhofer, G. (eds.) WG 1994. LNCS, vol. 903, pp. 151–163. Springer, Heidelberg (1995)
21. Wimmer, M., Kappel, G., Kusel, A., Retschitzegger, W., Schönböck, J., Schwinger, W., Kolovos, D., Paige, R., Lauder, M., Schürr, A., Wagelaar, D.: A Comparison of Rule Inheritance in Model-to-Model Transformation Languages. In: Cabot, J., Visser, E. (eds.) ICMT 2011. LNCS, vol. 6707, pp. 31–46. Springer, Heidelberg (2011)

Automatic Conformance Testing of Optimized Triple Graph Grammar Implementations*

Stephan Hildebrandt, Leen Lambers, Holger Giese,
Dominic Petrick, and Ingo Richter

Hasso Plattner Institute at the University of Potsdam
Prof.-Dr.-Helmert-Straße 2-3
14482 Potsdam, Germany
{stephan.hildebrandt,leen.lambers,holger.giese}@hpi.uni-potsdam.de,
{dominic.petrick,ingo.richter}@student.hpi.uni-potsdam.de

Abstract. In model-driven development, model transformations can be specified using an operational (imperative) or relational (declarative) approach. When using a relational approach, approved formal concepts are necessary to derive a conform operationalization. In general, though, it is not sure if implementations realize these formal concepts in an entirely correct way. Moreover, usually, available formal concepts neither cover every technicality, nor cover each additional optimization an implementation relies on. Consequently, conformance needs to be validated also on the implementation level. Conformance means that for each source model S and target model T related according to the relational specification, a corresponding implementation transforms S into T (and T into S in case that the specification is bidirectional).

We present an automatic conformance testing approach for TGG implementations, where the Triple Graph Grammar (TGG) approach is an important representative of relational model transformation approaches. We show that the grammar character of TGGs is very convenient for the automatic generation of conformance test cases. In particular, test input models can be generated together with their expected result obtaining a complete oracle. We show how to measure test suite quality and evaluate our approach on our own TGG implementation.

Keywords: Conformance testing, model transformation, triple graph grammar, relational specification, model-driven development.

1 Introduction and Motivation

When models and model transformations become most important artefacts as proposed in the MDE approach, it is essential to be able to rely on their correctness. In particular, model transformations holding errors may cause problems in

---

* This work was partially developed in the course of the project – Correct Model Transformations – Hasso Plattner Institut, Universität Potsdam and was published on its behalf and funded by the Deutsche Forschungsgemeinschaft. See http://www.hpi.uni-potsdam.de/giese/projekte/kormoran.html?L=1

A. Schürr, D. Varró, and G. Varró (Eds.): AGTIVE 2011, LNCS 7233, pp. 238–253, 2012.

the complete development chain. When using a relational model transformation approach, errors may arise because of faulty operationalizations. Therefore, in this paper, we concentrate on conformance testing for relational model transformation specifications and their implementations. *Conformance* means that for each source model S and target model T related according to the relational specification, a corresponding implementation transforms S into T (and T into S in case that the specification is bidirectional).

The *Triple Graph Grammar* (TGG) approach is an important representative of the *relational model transformation approach*. Basic theoretical concepts describing the derivation of TGG operationalizations are available ever since the TGG approach emerged [31]. To a certain extent, formal reasoning can be applied to prove conformance of the TGG and a corresponding operationalization [32,21,15]. In general, though, it is not sure if implementations realize each formal concept describing a conform operationalization in an entirely correct way. Moreover, usually, TGG formalizations neither cover every technicality TGG implementations rely on, nor cover each additional optimization augmenting efficiency of the model transformation execution. In practice, there exist different TGG implementations, realizing slightly different dialects, such as Fujaba TGG Engine [5], MOFLON [1], or ATOM3 [25]. Furthermore, even for a single tool holds that different tool versions with different optimizations exist. Consequently, in order to ensure the correct automation of model transformations, *conformance* needs to be validated also on the *implementation level*. In principle, back-to-back testing could be adopted to validate that different TGG implementations are equivalent, and regression testing could be adopted to validate that an implementation and its optimized version are conform. However, apart from that this may become very tedious, it does not ensure conformance with the specification. Therefore, in this paper we present an *automatic conformance testing* approach for TGGs and their implementations. We are confident that our approach is a good basis to develop conformance testing approaches also for other relational specification techniques such as QVT relational [29]. In particular, making use of [19] where an implementation of QVT relational with TGGs is presented.

For our *conformance testing approach* we exploit the following typical features of TGGs: (1) Because of its *grammar* character a TGG may serve as an "executable" model transformation contract being able to generate all correctly related input/output models obtaining a complete oracle. (2) A TGG describes *bidirectional* model transformations. We aim at providing the *TGG implementation developer* with automatic conformance testing support by generating test cases (exploiting the first feature) for forward as well as backward transformations (exploiting the second feature).

With conformance testing we mean that we search for errors in the implementation violating conformance with the specification [30] instead of finding errors in the specification itself. Moreover, we generate test cases that start with a "correct" input according to the specification, i.e., we present an approach for positive, not negative testing. Note that our test approach is designed to

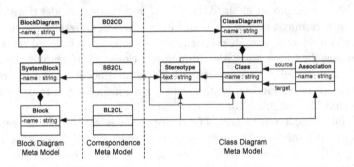

Fig. 1. Example metamodels

be applied to different versions or optimizations of the same TGG tool or even different TGG implementations, e.g., MOFLON [1] or the TGG Interpreter [19]. However, in the latter case, the problem of different input formats of *TGG Rules* poses an obstacle in practice. Different TGG implementations usually expect TGG rules in different technical formats. Therefore, a format conversion would be required before other TGG implementations can be tested with our framework. Furthermore, TGGs may specify non-deterministic forward and backward relations between source and target models. However, a model transformation is usually expected to return a deterministic result for a given input. Moreover, as explained in Sect. 3.3, having non-deterministic TGGs, it may not be possible to use our automatic test approach. Therefore, we impose some restrictions on the TGG rules (as explained more in detail in [15,16]) that can be verified statically to guarantee determinism.

Paper Outline. In Section 2, we introduce TGGs as relational model transformation specification technique and explain what it means for implementations to be conform with such a specification. We continue in Section 3 with a description of how to automate conformance testing of TGG implementations and how a corresponding test environment looks like. We also concentrate on measuring test suite quality. In Section 4, we show the results of testing our own TGG implementation and evaluate our approach. We give a description of related work in Section 5. Section 6 gives a summary and an overview on future work.

2   Conform Triple Graph Grammar Implementations

In general, three kinds of transformations can be performed with TGGs: Forward, backward, and correspondence transformations. A forward (backward) transformation takes a source (target) model as input and creates the correspondence and target (source) model. A correspondence transformation requires a source and target model and creates only the correspondence model. Subsequently, we concentrate on forward transformations. Analogous results can be derived for the backward case (which is symmetric) and the correspondence case.

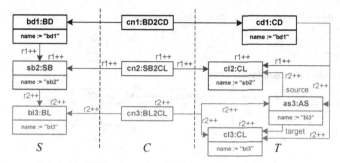

a++: Created by TGG axiom; r1++: TGG rule 1; r2++: TGG rule 2

Fig. 2. Forward transformation via the relational scheme

TGGs as Relational Specification Technique: To illustrate the following explanations, we will use a model transformation from simple SDL block diagrams[1] to UML class diagrams. The source and target metamodel of this model transformation are shown in Fig. 1. When using TGGs to specify model transformations, apart from the source and target metamodel, there is also a so-called correspondence metamodel. Its elements store traceability information, allowing to find elements from the source model that correspond to the target model, i.e., the correspondence model partly overlaps with the source and target models. TGGs relate three different models: A source model, a target model, and a correspondence model connecting the source and target model to a so-called triple graph. Further on, we use a triple of variables SCT to denote one triple graph, where S denotes its source component, C its correspondence component, and T its target component. In our example, block and class diagrams are the source and target graphs connected by a correspondence graph, constituting a triple graph (see Fig. 2). A TGG consists of an axiom $S_A C_A T_A$ (the grammar's start graph)[2] and several TGG rules that are always creating.[3] The TGG for the transformation of block and class diagrams is shown in Fig. 3.[4] Elements that are preserved are drawn black, elements that are created are drawn green and marked with "++". The TGG can be used to build triple graphs, representing correctly related source and target models, as follows: Starting from the axiom, the rules of the TGG are applied wherever they match in an arbitrary order leading to a derivation tree like shown in Fig. 4. Each arrow represents a

---

[1] This is a simplified version of SDL block diagrams
(http://www.itu.int/ITU-T/studygroups/com17/languages/Z100.eps)

[2] In particular, we use a so-called axiom rule, which is applied once to the empty graph and thereby sets some attribute values creating the concrete axiom.

[3] We restrict our explanations to TGG rules without application conditions, but it is possible to apply our test framework also to more expressive TGGs holding application conditions[22,21,19]. In our TGG implementation it is possible to express application conditions using OCL constraints.

[4] Note, that the types defined in Fig. 1 are abbreviated in Fig. 3.

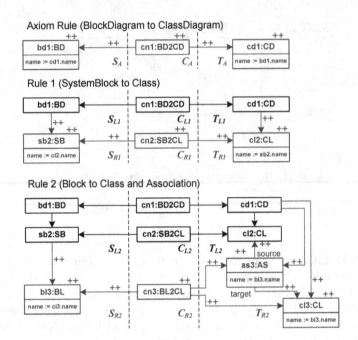

Fig. 3. Example TGG rules and axiom rule

rule application and each node represents a triple graph of the so-called *TGG language* $\mathcal{L}(TGG)$ containing all correctly related models according to the TGG. Applying TGG rules in this manner is used later on to automatically generate source and target test models (see Sect. 3).

The most straightforward way to derive a forward translation from the relational TGG follows the so-called *relational scheme*: Given a source graph S, then the forward translation via the relational scheme returns a set of triple graphs that can be generated by the TGG starting from $S_A C_A T_A$ and have S as a source component. Fig. 2 shows a block diagram S with a corresponding class diagram T connected by a correspondence model. Assuming that we want to transform the depicted block diagram S, T is a valid forward transformation result following the relational scheme because the triple graph SCT is created according to the TGG in Fig. 3. The annotations in Fig. 2 indicate which rules create the corresponding elements. The attribute assignments ensure equality of the name attributes of corresponding elements. However, while all three models are created in parallel, the actual name values can be chosen arbitrarily. Note that the generation of SCT corresponds conceptually to one path in the derivation tree of Fig. 4.

Conform TGG Implementation: In practice, when performing a forward transformation via some TGG, one of the models already exists and a model transformation system has to create the other one. Therefore, operational rules have to be generated from the TGG, which create consistently related target

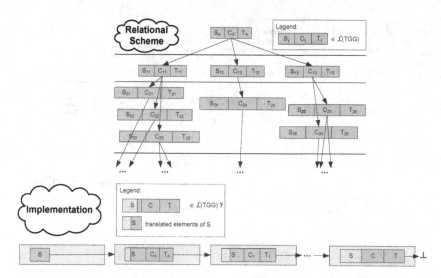

Fig. 4. Relational scheme and implementation

model elements for given source model elements. For each of the aforementioned transformation directions, separate operational rules can be derived and they are applied to the model to be translated in some specific optimal way. This results in a linear derivation for the implementation like shown in Fig. 4. Thereby, in order to ensure conformance with the relational scheme, it is important that the resulting triple graph belongs to the aforementioned triple graph language. For example, Fig. 2 shows a forward translation following the relational scheme from a block diagram into a class diagram. A conform implementation would start its translation with the *BlockDiagram* and transforms it into the depicted *ClassDiagram*. The following section is concerned with testing this kind of conformance.

Our TGG Implementation: We have developed an implementation[5] of TGGs based on Eclipse and the Eclipse Modeling Framework.[6] The system can perform model transformations (see [15,16] for more information and formal details) and model synchronizations[18]. It utilizes several optimizations to increase performance of model transformations, for example, limiting the pattern matching process to parts of the source model. Fig. 5 shows an overview of the tool's architecture and the test framework, which will be presented in Sect. 3. The core component of our tool architecture is the *TGG Engine*, which performs model transformations specified by a set of *TGG Rules*, which can be edited by a *TGG Editor*. While TGGs are a purely relational formalism, an operational form has to be derived to make them executable. In our implementation, a set of *Story Diagrams* [11] is generated for each TGG rule by an *Operational Rules Generator*. *Story Diagrams* combine UML activity diagrams to express control flow, and graph transformation rules to express pattern matching and modifications

---

[5] Downloadable from our Eclipse update site http://www.mdelab.de/update-site
[6] http://www.eclipse.org/emf

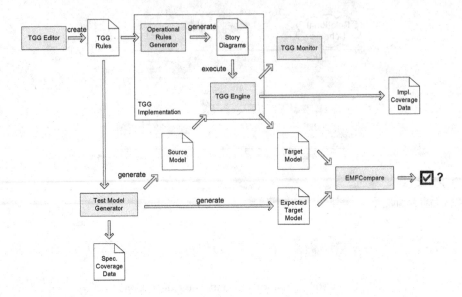

Fig. 5. Test environment

on graphs. For each TGG rule, three story diagrams are generated, one for each transformation direction. These *Story Diagrams* are executed by a *Story Diagram Interpreter* [17]. The *TGG Engine* invokes the interpreter to execute the appropriate *Story Diagrams*.

3   Automatic Conformance Testing of TGG Implementations

3.1   Test Objective

As explained in Sect. 2, the TGG language $\mathcal{L}(TGG)$ contains all correctly related source and target models SCT according to the TGG. Our *test objective* is checking conformance of the TGG implementation with the TGG. A TGG and its implementation are *conform* if for each SCT belonging to $\mathcal{L}(TGG)$ it holds that S is translated forward into T and T is translated backward into S by the TGG implementation. In particular, we want to analyze that for each node $S_{ij}C_{ij}T_{ij}$ in Fig. 4, it holds that S_{ij} is translated forward into T_{ij} by the implementation and T_{ij} is translated backward into S_{ij}.[7]

In our implementation, possible sources of errors are the *TGG Engine* and the *Operational Rules Generator* because these components make up the *TGG Implementation*. Causes may originate from an implementation that does not realize each formal concept describing a conform operationalization in an entirely

---

[7] In principle, one could also include the correspondence component into the test objective. However, we ignore the correspondences here, because we are mainly interested in the model transformation result T.

correct way. Moreover, an implementation might include erroneous optimizations, or domain-specific technicalities not considered in the TGG formalization so far. In Sect. 4, we present such technical specialties.

3.2 Test Case Definition and Random Generation

Given the test objective, we can define a *test case* for the forward case (backward analogous) as follows: it consists of a source graph S and the expected target graph T such that there exists a triple graph SCT in the TGG language $\mathcal{L}(TGG)$.

Usually, one major problem when generating test cases for model transformations is how to obtain the expected target models as test oracles (see Sect. 5 on related work). However, in the case of TGGs, their grammar character (see Sect. 2 and Fig. 4) allows us to automatically generate triple graphs SCT in a random way, representing a test case input with a source graph S and expected target graph T. This *random generation* procedure is implemented by the *Test Model Generator* (cf. Fig. 5). First, it creates the elements of the axiom (the grammar's start graph). Then it randomly applies all TGG rules a predefined number of times to extend the start graph. At this point, the source S and target model T are extracted from the triple graph SCT.

3.3 Test Case Execution

A test case consists of a test case input S and the expected target graph T, generated randomly as described in the last section. The source model S is transformed by the TGG implementation under test to create a second target model T'. According to the test objective, the expected target model T and T' must be equal to pass the test successfully, otherwise one has detected a conformance error. Here, we exploit the fact that the model transformation is deterministic (cf. Sect. 1), i.e., there is only one target graph T corresponding to S. Otherwise, we would have to generate the set of all target models to a specific source model, and check whether the created target model T' is contained in that set. This would greatly increase computation effort if it is at all possible (e.g., the number of possible target models could be infinite).

Technically, we compare the expected target model T and computed model T' using *EMFCompare*. It uses a heuristic to find pairs of corresponding elements in both models. All elements of the two models are compared by a similarity metric based on types, values of attributes and references of both elements. *EMFCompare* is able to detect identical models based on this metric. Moreover, it outputs information about differences between both models. More information can be found in [34]. Note that although this technical solution is quite satisfactory already, model comparison is a research topic on its own as described in [26].

If a conformance error has been detected, we need some means to find its cause. To this extent, we provide a *TGG Monitor* logging the internal state of the *TGG Engine* during a model transformation execution allowing the developer to inspect the model transformation step-by-step.

3.4   Test Suite Quality

The grammar character of TGGs gives us a means to randomly generate test cases. However usually, an infinite number of them exists so we have to limit the size of test cases. Therefore, we need some well-considered techniques to measure the quality of a test suite, consisting of a finite set of test cases. To this extent, we apply different test adequacy criteria. We not only consider the TGG as an oracle, but we consider it also to develop specification coverage criteria. Moreover, we evaluate the quality of our test suite with respect to implementation coverage.

Specification coverage is analyzed by checking for a given test suite which TGG rules it covers. In particular, if each TGG rule was used at least once[8] to generate test case inputs and corresponding expected results of the test suite, then the correct implementation of each TGG rule is tested. Therefore, we have the following *TGG rule coverage criterion*: Given a test suite, each TGG rule has been applied at least once in order to generate the test suite. TGG rule coverage can be measured by evaluating test cases that were randomly generated (see Sect. 3.2). During random generation of test cases, the *Test Model Generator* (see Fig. 5) records, which of the TGG rules have been applied such that it can provide us also with coverage data.

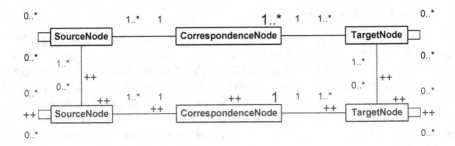

Fig. 6. Prototypical structure of TGG rules in our implementation

As mentioned in [20,24] errors in a rule-based implementation often arise because of dependencies. *Rule dependency coverage* then describes how many of these dependencies are covered by test suite. In particular, TGG rules might create relations between source and target elements depending on the existence of another relation between specific source and target elements (these rules are said to be in a produce-use relation accordingly). For the corresponding forward rules in the implementation, this means that specific source elements should be translated only if some related source elements have been translated before already. For example, rule 2 in Fig. 3 depends on the axiom and rule 1. It is possible to find out these dependencies statically on specification level using AGG [4] and systematically generate test cases covering them. For efficiency

---

[8] In principle, TGGs exist in which specific rules are never applicable. For these kind of inconsistent TGGs the rule coverage criterion will never be fulfilled.

reasons, we implemented a simple dependency detection algorithm that exploits certain TGG rule properties specific to our TGG implementation summarized by the prototypical structure shown in Fig. 6. Amongst other criteria, this prototypical structure is required to ensure deterministic transformations [15] and, moreover, allows us to implement specific optimizations. Due to the fact that all model elements are connected to a single correspondence node, we can compute dependencies based on the produce-use relation between correspondence node types only. This leads to an over-approximation because dependencies might be detected that do not really exist.

Implementation coverage considers the concrete model transformation implementation of the TGG and computes for a given set of test cases which parts of the implementation have been covered when executing the test cases. This information helps to find out, whether all relevant parts of the TGG implementation are covered by the test cases. Our implementation (see Fig. 5) is model-based (story diagrams) and we have a TGG Engine triggering the interpretation of these models by surrounding code. Therefore, we consider two kinds of coverage: *Story diagram coverage* and *code coverage*.

Story Diagram Coverage: The story diagrams generated by the *Operational Rules Generator* contain large parts of the model transformation algorithm, e.g., for bookkeeping of transformed elements, detecting inconsistencies in the input models, optimizing rule application. A detailed measurement of story diagram coverage shows whether a test case covers all relevant parts of the story diagrams. Some coverage criteria for Activity Diagrams are described by Chen et al. [7], e.g., activity coverage, which is the ratio between checked activities and all activities in the activity diagram, and transition coverage, which is the ratio between checked transitions and all transitions in the activity diagram. Story diagram coverage as used here is the ratio between all executed elements (i.e., activities and transitions as well as nodes and links of graph transformation rules) and the total number of elements in all executed story diagrams of a transformation direction.

Code Coverage: Elaborate code coverage techniques have been developed already in the past decades. We use the code coverage tool EclEMMA[9] to measure statement coverage of the TGG Engine. Here, we cannot expect full code coverage, since the TGG Engine contains code, e.g., for executing model transformations via the GUI, or executing a specific transformation direction, respectively.

4 Testing Our TGG Implementation and Evaluation

We have executed conformance test cases for *three different TGGs*, which represent typical uses of model transformations: Model-to-model transformations (SDL2UML), model-to-text transformations (Automata2PLC), and a large practical model transformation from an industrial project (SystemDesk2AUTOSAR). The SDL2UML TGG is a slightly more complex version of the TGG presented in

---

[9] http://www.eclemma.org/

Fig. 3. The TGG contains six rules, where the largest rules contain 17 nodes. Automata2PLC [14] is a transformation from automata models to abstract syntax trees (AST) of a language for programmable logic controllers (PLC). It consists of five rules with up to 20 nodes in a rule. In contrast to the other two transformations, which are bidirectional, this transformation can only transform from automata to PLC models but not reverse. The SystemDesk2AUTOSAR transformation [18] was created in an industrial research project. It transforms models from the modeling tool dSPACE SystemDesk to AUTOSAR, a standardized modeling language in the automotive domain. This transformation consists of 50 rules and the largest of them contain 31 nodes.

Test cases were generated with 10, 100, and 1000 random rule applications. TGG rules are creating, therefore the test model size correlates to the number of rule applications. For example, the rules of SDL2UML TGG create one element in the SDL model and five elements in the UML model on average. Therefore, 1000 random rule applications of that TGG lead to an SDL model with 1000 elements and a UML model with about 5000 elements. Five test cases were generated for each of the three numbers of rule applications, each TGG, and each possible transformation direction. From each group of five test cases the test case with the highest rule coverage was selected for further execution. This amounts to a total of 15 executed test cases. Executing a test case with 1000 rule applications takes about two minutes on an Intel i5 750 CPU with 2.67 GHz. The smaller test cases take less then a minute. For the two *smaller TGGs*, we could easily achieve *complete rule* and *dependency coverage*, even for the smallest test models. However, for the *very large* SystemDesk2AUTOSAR TGG, we could *not* achieve *complete* specification coverage. *Rule coverage* for test models with 10 rule applications was only 37% and went up to 71% (both directions) for the test models with 1000 rule applications. *Dependency coverage* was only 19% for the largest test models. The SystemDesk2AUTOSAR TGG contains many rules with quite complex preconditions. For some of them, it is rather improbable to create a triple graph satisfying the required precondition with a purely random approach. Therefore, specification coverage increases with a higher number of rule applications, but is still far from complete coverage. Here, a *systematic approach* (possibly making use of model checking [13]) to generate test models would be required.

For the SDL2UML TGG, we achieved a *story diagram coverage* of 82% for the largest test models. Story diagram coverage of the Automata2PLC TGG is 97% for all test models. Like specification coverage, story diagram coverage of the SystemDesk2AUTOSAR TGG was rather low, only 36% for the largest test models. Naturally, story diagram coverage is *connected to specification coverage*, because the story diagrams can only be completely executed if specification coverage is complete. In addition, the generated story diagrams contain code for error handling, which is never executed in the test cases. Therefore, story diagram coverage of the SDL2UML and Automata2PLC TGGs can only be improved with test cases using faulty input models (negative testing).

Code coverage of the TGG Engine (49%) is quite *constant* over all test cases. When looking into detail, we can see that all *code parts* of the TGG Engine that are *relevant* to the test cases are *covered*.

The *results* of the test cases indicate, that the SDL2UML TGG and its implementation are indeed conform, because EMFCompare did not find any differences between transformed and expected target models. The test cases for the other two TGGs failed indicating that complete conformance with their implementations could not be shown, mainly because of *domain-specific technicalities not covered by TGG formalizations so far*. Until now, we did neither find errors related to the faulty interpretation of formal concepts for TGG operationalizations nor for optimizations augmenting efficiency of the model transformation execution.

In the Automata2PLC test case, we discovered that the transformation does not maintain the *order of model elements* in references. The *TGG Implementation* creates target elements in a different order than the *Test Model Generator*, which is reported by EMFCompare. This is not a problem for models without ordered references. However, in the PLC target model all references are ordered references. The PLC model is an abstract-syntax tree of a textual language. The order of the elements in the syntax tree directly reflects the order of the corresponding text blocks of the textual syntax. Therefore, this order is relevant like in most other textual languages. The Automata input model does not have ordered references. It consists of states connected by transitions. For the PLC model, we have to derive an order in some way. However, there is no formalization of TGGs taking ordered references into account.

The SystemDesk2AUTOSAR test cases revealed another subtle problem in the forward and backward directions. For example, SystemDesk models contain a *Library* element, which does not have a counterpart in AUTOSAR (this problem occurs in several places, not just with *Libraries*, and it also occurs in the opposite direction, i.e., there are elements in AUTOSAR that do not have a direct counterpart in SystemDesk). While all elements in SystemDesk must have a UUID (universally unique identifier), each *Library* gets a new random UUID when it is created. Of course, the *Library* in the expected target model has a different UUID than the *Library* element in the transformed target model. Therefore, EMFCompare reports a difference between both models, although they are otherwise identical. One can argue that this transformation is not deterministic because it creates random UUIDs in the target model. One idea to overcome this problem would be to assign a static UUID for the *Library* in the TGG rule that creates it. However, this is not an option in practice because then the *Library*'s UUID would not be *universally unique* anymore, which would cause problems in the modeling tools. Therefore, another idea is to derive the *Library*'s UUID from the source model in some way to achieve determinism. However, a proper way to do that has yet to be defined. Another possibility is to adapt EMFCompare so that it ignores UUIDs for *Library* (and other affected) elements. However, this adaptation of the test framework would be specific to these models. Summarizing, the concepts of UUIDs as well as ordered references should be considered by TGG formalizations such that conform TGG implementations can be built properly. Existing TGG formalizations are defined on attributed graphs. On the contrary, models are based on metamodeling standards like MOF or EMF including additional concepts specific to the technical modeling domain.

Finally, we still have the following *limitation* of the test framework: The example TGGs that we tested only contain string attributes without complex attribute computations. During generation of the input and expected target models, these attributes are set to arbitrary values. For complex data types and attribute computations, more complex techniques like classical data partitioning techniques have to be integrated with the *Test Model Generator*. Especially, if the attributes' values have to satisfy certain constraints.

5   Related Work

Compiler testing as oldest line of related work is based on the grammar-based generation of test data. [23] gives a survey on compiler testing methods and presents a number of coverage criteria. The rule coverage criterion is the most basic one and we use this coverage criterion as an orientation in our grammar-based generation of input test models for TGG implementations (see Section 3).

There is some work on validating and verifying relational (or declarative) model transformation specifications themselves (e.g., [6,27]). In this work, we assume that the specification has been validated and aim at testing conformance of the specification with its implementation.

A number of conformance testing approaches relying on graph transformation as a specification technique exist [9,20,2]. Instead of focusing on model transformation specifications and implementations, they are rather concerned with conformance testing of behavioral specifications w.r.t. (actual) behavior in refined models or (generated) code.

There are a number of testing approaches proposed for model transformation implementations. Most black-box methods are concerned with generating (as e.g., [33,8,12]) qualified test input models taking into consideration the input metamodel (and corresponding constraints). For example, in [12] meta-model coverage is considered using data-partitioning techniques. It is required, for example, that models must contain representatives of association ends, which differ in their cardinalities. On the contrary, [24] proposes white-box criteria to qualify test input models. We concentrate on generating conformance test cases using the model transformation specification as an "executable contract" generating not only test input models, but also expected results obtaining a complete oracle. [3,28] mention that, in general, describing the oracle is a difficult task because even simplest expected results may become quite complex. The availability of formal requirements for the model transformation is desirable and can be used for building the oracle. In particular, [10] presents an approach for specifying MOF-based metamodels and their interrelationships, model transformation specification, implementation and test case generation using constructive logic. [10] uses the specification as partial oracle and does not generate expected results as we do. Moreover, it proposes a new uniform framework, whereas we rely on TGGs as existing model transformation specification technique for which several tools are already available.

6   Conclusion and Future Work

We have presented and evaluated an approach to automatic conformance testing for optimized TGG implementations. The grammar character of TGGs is used to generate test cases including expected results. We are confident that our approach is a good basis to develop automatic conformance testing also for other relational specification techniques.

As future work we consider a number of extensions to this approach. (1) Our TGG implementation supports not only batch model transformations, but also model synchronization. We would like to develop an automatic conformance test approach also for the latter case. (2) Since we concentrate on positive testing, the approach should be extended to negative testing to find out if a TGG implementation handles inconsistent input models in a reasonable way. This is not trivial, since parsing would be involved to find out if a model would be an inconsistent input according to the TGG. (3) As in the approaches for generating test input models based on the source metamodel of a model transformation, it would be interesting to integrate metamodel constraints. Certain models consistent according to the TGG are potentially inconsistent w.r.t. the metamodel constraints and become negative test input models. (4) We have described first test case generation strategies and coverage criteria. We plan to apply mutation testing or more systematic generation strategies (possibly making use of model checking) to qualify test cases. (5) As mentioned in the introduction, our aim is to generalize this automatic test approach to other relational model transformation approaches such as QVT Relational.

Acknowledgement. We thank the anonymous reviewers for their valuable comments.

References

1. Amelunxen, C., Klar, F., Königs, A., Rötschke, T., Schürr, A.: Metamodel-based tool integration with MOFLON. In: ICSE 2008, pp. 807–810. ACM Press (2008)
2. Baldan, P., König, B., Stürmer, I.: Generating Test Cases for Code Generators by Unfolding Graph Transformation Systems. In: Ehrig, H., Engels, G., Parisi-Presicce, F., Rozenberg, G. (eds.) ICGT 2004. LNCS, vol. 3256, pp. 194–209. Springer, Heidelberg (2004)
3. Baudry, B., Ghosh, S., Fleurey, F., France, R., Le Traon, Y., Mottu, J.M.: Barriers to systematic model transformation testing. Communications of the ACM 53, 139–143 (2010)
4. Biermann, E., Ermel, C., Lambers, L., Prange, U., Runge, O., Taentzer, G.: Introduction to AGG and EMF Tiger by modeling a conference scheduling system. International Journal on Software Tools for Technology Transfer 12(3-4) (2010)

5. Burmester, S., Giese, H., Niere, J., Tichy, M., Wadsack, J.P., Wagner, R., Wende-hals, L., Zündorf, A.: Tool integration at the meta-model level within the FUJABA Tool Suite. International Journal on Software Tools for Technology Transfer 6(3), 203–218 (2004)

6. Cabot, J., Clarisó, R., Guerra, E., de Lara, J.: Verification and validation of declarative model-to-model transformations through invariants. Journal of Systems and Software 83(2), 283–302 (2010)

7. Chen, M., Mishra, P., Kalita, D.: Coverage-driven automatic test generation for UML Activity Diagrams. In: Proceedings of the 18th ACM Great Lakes Symposium on VLSI, pp. 139–142. ACM (2008)

8. Ehrig, K., Küster, J.M., Taentzer, G., Winkelmann, J.: Generating Instance Models from Meta Models. In: Gorrieri, R., Wehrheim, H. (eds.) FMOODS 2006. LNCS, vol. 4037, pp. 156–170. Springer, Heidelberg (2006)

9. Engels, G., Güldali, B., Lohmann, M.: Towards Model-Driven Unit Testing. In: Kühne, T. (ed.) MoDELS 2006. LNCS, vol. 4364, pp. 182–192. Springer, Heidelberg (2007)

10. Fiorentini, C., Momigliano, A., Ornaghi, M., Poernomo, I.: A Constructive Approach to Testing Model Transformations. In: Tratt, L., Gogolla, M. (eds.) ICMT 2010. LNCS, vol. 6142, pp. 77–92. Springer, Heidelberg (2010)

11. Fischer, T., Niere, J., Torunski, L., Zündorf, A.: Story Diagrams: A New Graph Rewrite Language Based on the Unified Modeling Language and Java. In: Ehrig, H., Engels, G., Kreowski, H.-J., Rozenberg, G. (eds.) TAGT 1998. LNCS, vol. 1764, pp. 296–309. Springer, Heidelberg (2000)

12. Fleurey, F., Steel, J., Baudry, B.: Validation in model-driven engineering: Testing model transformations. In: First International Workshop on Model, Design and Validation, pp. 29–40. IEEE Computer Society (2004)

13. Fraser, G., Wotawa, F., Ammann, P.E.: Testing with model checkers: A survey. Software Testing, Verification and Reliability 19(3), 215–261 (2009)

14. Giese, H., Glesner, S., Leitner, J., Schäfer, W., Wagner, R.: Towards verified model to code transformations. In: Proc. of the 3rd Workshop on Model Design and Validation. ACM/IEEE, Genova, Italy (2006)

15. Giese, H., Hildebrandt, S., Lambers, L.: Toward bridging the gap between formal semantics and implementation of triple graph grammars. Tech. Rep. 37, Hasso Plattner Institute at the University of Potsdam (2010)

16. Giese, H., Hildebrandt, S., Lambers, L.: Toward bridging the gap between formal semantics and implementation of triple graph grammars. In: Lúcio, L., Vieira, E., Weißleder, S. (eds.) Proceedings of Models Workshop on Model-Driven Engineering, Verification and Validations, pp. 19–24. IEEE Computer Society (2010)

17. Giese, H., Hildebrandt, S., Seibel, A.: Improved flexibility and scalability by interpreting story diagrams. In: Magaria, T., Padberg, J., Taentzer, G. (eds.) Proceedings of the 8th International Workshop on Graph Transformation and Visual Modeling Techniques. Electronic Communications of the EASST, vol. 18 (2009)

18. Giese, H., Hildebrandt, S., Neumann, S.: Model Synchronization at Work: Keeping SysML and AUTOSAR Models Consistent. In: Engels, G., Lewerentz, C., Schäfer, W., Schürr, A., Westfechtel, B. (eds.) Graph Transformations and Model-Driven Engineering. LNCS, vol. 5765, pp. 555–579. Springer, Heidelberg (2010)

19. Greenyer, J., Kindler, E.: Comparing relational model transformation technologies: Implementing Query/View/Transformation with triple graph grammars. Software and Systems Modeling 9(1), 21–46 (2010)

20. Heckel, R., Mariani, L.: Automatic Conformance Testing of Web Services. In: Cerioli, M. (ed.) FASE 2005. LNCS, vol. 3442, pp. 34–48. Springer, Heidelberg (2005)

21. Hermann, F., Ehrig, H., Golas, U., Orejas, F.: Efficient analysis and execution of correct and complete model transformations based on triple graph grammars. In: Proceedings of the 1st International Workshop on Model-Driven Interoperability, pp. 22–31. ACM (2010)
22. Kindler, E., Wagner, R.: Triple graph grammars: Concept, extensions, implementations and application scenarios. Tech. rep., Software Engineering Group, Department of Computer Science, Universität Paderborn (2007)
23. Kossatchev, A.S., Posypkin, M.A.: Survey of compiler testing methods. Programming and Computer Software 31, 10–19 (2005)
24. Küster, J.M., Abd-El-Razik, M.: Validation of Model Transformations – First Experiences Using a White Box Approach. In: Kühne, T. (ed.) MoDELS 2006. LNCS, vol. 4364, pp. 193–204. Springer, Heidelberg (2007)
25. de Lara, J., Vangheluwe, H.: Using AToM3 as a Meta-CASE environment. In: International Conference on Enterprise Integration Systems (2002)
26. Lin, Y., Zhang, J., Gray, J.: Model comparison: A key challenge for transformation testing and version control in model driven software development. In: Control in Model Driven Software Development. OOPSLA/GPCE: Best Practices for Model-Driven Software Development, pp. 219–236. Springer (2004)
27. Lin, Y., Zhang, J., Gray, J.: A testing framework for model transformations. In: Beydeda, S., Book, M., Gruhn, V. (eds.) Model-Driven Software Development, pp. 219–236. Springer (2005)
28. Mottu, J.M., Baudry, B., Traon, Y.L.: Model transformation testing: Oracle issue. In: Proceedings of the 2008 IEEE International Conference on Software Testing Verification and Validation Workshop, pp. 105–112. IEEE Computer Society (2008)
29. Object Management Group: MOF 2.0 QVT 1.0 Specification (2008)
30. Richardson, D., O'Malley, O., Tittle, C.: Approaches to specification-based testing. In: Kemmerer, R. (ed.) Proc. of the ACM SIGSOFT 1989 Third Symposium on Software Testing, Analysis, and Verification, pp. 86–96. ACM Press (1989)
31. Schürr, A.: Specification of Graph Translators with Triple Graph Grammars. In: Mayr, E.W., Schmidt, G., Tinhofer, G. (eds.) WG 1994. LNCS, vol. 903, pp. 151–163. Springer, Heidelberg (1995)
32. Schürr, A., Klar, F.: 15 years of Triple Graph Grammar: Research Challenges, New Contributions, Open Problems. In: Ehrig, H., Heckel, R., Rozenberg, G., Taentzer, G. (eds.) ICGT 2008. LNCS, vol. 5214, pp. 411–425. Springer, Heidelberg (2008)
33. Sen, S., Baudry, B., Mottu, J.-M.: Automatic Model Generation Strategies for Model Transformation Testing. In: Paige, R. (ed.) ICMT 2009. LNCS, vol. 5563, pp. 148–164. Springer, Heidelberg (2009)
34. Toulmé, A.: Presentation of EMF Compare utility. In: Eclipse Modeling Symposium 2006, pp. 1–8 (2006), http://www.eclipsecon.org/summiteurope2006/index.php?page=detail/&id=6

Author Index